Design Automation of Quantum Computers

Rasit O. Topaloglu
Editor

Design Automation
of Quantum Computers

 Springer

Editor
Rasit O. Topaloglu
IBM
Poughkeepsie, NY, USA

ISBN 978-3-031-15701-1   ISBN 978-3-031-15699-1   (eBook)
https://doi.org/10.1007/978-3-031-15699-1

This Springer imprint is published by the registered company Springer Nature Switzerland AG
The registered company address is: Gewerbestrasse 11, 6330 Cham, Switzerland

Preface

Quantum computers are on the cusp of having navigated the wavy roads from theory to practice. I remember vividly that a decade ago, less than a handful of qubits on hardware was boasted as the state of the art and yet generated awe in me and others. Fast forward to today, there are quantum computers that have shattered the hundred qubit ceiling and the next big mark is a thousand qubits.

Beyond hundred qubits, manual process of the design flow starts to become a bottleneck. One of the voids that this book targets is to provide means for designing quantum computers with hundred qubits or more. In the first chapter, researchers from Johannes Kepler University Linz introduce decision diagrams as a method to cope with memory limitations for simulation and verification of quantum systems. Second chapter by UCLA targets the problem of optimally mapping logical qubits to physical ones on the hardware. Third chapter from UCSB is a timely treatment of quantum processor architecture design.

Beyond the simulation, verification, and design aspects of quantum computers, there are also application aspects. In the fourth chapter, Penn State researchers utilize a quantum computer to generate random numbers, which have many applications including cryptography.

One branch of quantum technology has been focusing on exploiting superconducting device speeds while utilizing traditional digital circuit logic. This branch uses flux-based devices. In the fifth chapter, National Tsing Hua researchers develop a placement algorithm for a single-flux quantum-based field programmable gate array. The next chapter by USC authors targets yield optimization aspects of the single-flux quantum technology. The book finishes off with an exploration of the single-flux quantum technology for hardware security, in particular logic locking aspects by University of Rochester researchers.

Just like that last talk before lunch, this is the last paragraph before leaving you with appetizing and yet easy-to-digest content from top academics. Thinking back, perhaps it was that awe that motivated me to learn more about this field and make contributions. With regard to this book, it has been not only enjoyable to read their chapters but also to work with these elite authors for improvements. I anticipate the contents of this book could indeed lead the way to advance the quantum field in terms of design automation of quantum computers.

To a quantum future!

Poughkeepsie, NY, USA   Rasit O. Topaloglu

Contents

Decision Diagrams for Quantum Computing 1
Robert Wille, Stefan Hillmich, and Lukas Burgholzer

Layout Synthesis for Near-Term Quantum Computing:
Gap Analysis and Optimal Solution 25
Bochen Tan and Jason Cong

Towards Efficient Superconducting Quantum Processor
Architecture Design 41
Gushu Li, Yufei Ding, and Yuan Xie

Quantum True Random Number Generator 69
Abdullah Ash Saki, Mahabubul Alam, and Swaroop Ghosh

Placement Algorithm of Superconducting Energy-Efficient
Magnetic FPGA 87
Sagar Vayalapalli, Yi-Chen Chang, Naveen Katam, and Tsung-Yi Ho

Margin Optimization of Single Flux Quantum Logic Cells 105
Mustafa Altay Karamuftuoglu, Soheil Nazar Shahsavani,
and Massoud Pedram

Hardware Security of SFQ Circuits 135
Tahereh Jabbari, Yerzhan Mustafa, Eby G. Friedman, and Selçuk Köse

Index 167

Decision Diagrams for Quantum Computing

Robert Wille (iD) **, Stefan Hillmich** (iD) **, and Lukas Burgholzer** (iD)

1 Introduction

Quantum computing promises to speed up many important applications even in the current NISQ era [1] and more so once fault-tolerance is achieved. The underlying primitives of quantum computing are fundamentally different to conventional computations. This introduces new challenges for design automation and software development such as the exponential memory requirement to store arbitrary quantum states and operations on non-quantum hardware.

The design automation community in the conventional domain has spent decades to successfully solve many difficult problems. One of these solutions which especially addresses memory consumption is the usage of decision diagrams to represent information. In the conventional domain, there exists a plethora of different types such as *Binary Decision Diagrams* (BDDs, [2]), *Binary Moment Diagrams* (BMDs, [3]), *Zero-suppressed Decision Diagrams* (ZDDs, [4]), or *Tagged BDDs* [5]. Inspired by the results achieved with decision diagram in the conventional domain, several types have been invented for the quantum domain, such as *X-decomposition Quantum Decision Diagrams* (XQDDs, [6]), *Quantum Decision Diagrams* (QDDs, [7]), *Quantum Information Decision Diagrams* (QuIDDs, [8]), or *Quantum Multiple-valued Decision Diagrams* (QMDDs, [9, 10]). However, many researchers and engineers working in the domain of quantum computing are still

R. Wille (✉)
Chair for Design Automation, Technical University of Munich, Munich, Germany
Software Competence Center Hagenberg GmbH (SCCH), Linz, Austria
e-mail: robert.wille@tum.de

S. Hillmich · L. Burgholzer
Institute for Integrated Circuits, Johannes Kepler University Linz, Linz, Austria
e-mail: stefan.hillmich@jku.at; lukas.burgholzer@jku.at

© The Author(s), under exclusive license to Springer Nature Switzerland AG 2023
R. O. Topaloglu (ed.), *Design Automation of Quantum Computers*,
https://doi.org/10.1007/978-3-031-15699-1_1

rather unfamiliar with the concepts of decision diagrams and, hence, often cannot fully exploit this potential.

In this chapter, we review decision diagrams as data structure to compactly represent quantum states and quantum operators. To this end, we explain how decision diagrams are obtained from decomposing state vectors along with an explanation of the graphical notation. The vector decomposition is subsequently extended to obtain decision diagrams for matrices. Afterwards, we cover selected applications of decision diagrams for design and validation work. More precisely, we begin by covering error-free quantum circuit simulation, which is essentially matrix-vector multiplication. In the next step, we discuss noisy quantum circuit simulation and the advantages decision diagrams have in this application. Additionally to simulation, we present as a main aspect of quantum circuit verification an efficient procedure to check the equivalence of quantum circuits using decision diagrams.

The remainder of this chapter is structured as follows. Section 2 gives the background on decision diagrams, specifically how vectors and matrices are represented. In Sects. 3 and 4, we show how decision diagrams can be employed to conduct quantum circuit simulation without and with noise, respectively. Section 5 explains how decision diagrams lead to more efficient equivalence checking procedures. Finally, Sect. 6 concludes the chapter.

2   Decision Diagrams

In this section, we describe how decision diagrams exploit redundancies in vectors and matrices to enable a compact representation in many cases. More precisely, we first detail the representation for state vectors which we subsequently extend by a second dimension to compactly represent matrices for quantum operations.

2.1   Representation of State Vectors

The representation of a system composed of n qubits on non-quantum hardware is commonly achieved through 2^n-dimensional vector—an exponential representation. However, a closer look at state vectors unveils that they are frequently composed of redundant entries which provide potential for a more compact representation.

Example 1 Consider a quantum system with $n = 3$ qubits situated in a state given by the following vector:

$$\psi = \left[0, 0, \tfrac{1}{2}, 0, \tfrac{1}{2}, 0, -\tfrac{1}{\sqrt{2}}, 0\right]^T.$$

Although exponential in size ($2^3 = 8$ entries), this vector only assumes three different values, namely 0, $\frac{1}{2}$, and $-\frac{1}{\sqrt{2}}$.

Redundancy in the considered data can be exploited to attain a compact representation. To this end, we propose to employ decision diagrams. For conventional computations, decision diagrams such as the *Binary Decision Diagram* (BDD, [2]) are a tried and tested data structure and have been used for decades. For BDDs, a decomposition scheme is employed which reduces a function to be represented into corresponding sub-functions. Since the sub-functions usually include redundancies as well, equivalent sub-functions result which can be shared—eventually yielding a much more compact representation. In a similar fashion, the concept of decomposition can also be applied to represent state vectors in a more compact fashion.

Similar to decomposing a function into sub-functions, we decompose a given state vector with its complex entries into sub-vectors. To this end, consider a quantum system with qubits $q_{n-1}, q_1, \ldots q_0$, whereby q_{n-1} represents the most significant qubit.[1] Then, the first 2^{n-1} entries of the corresponding state vector represent the amplitudes for the basis states with q_{n-1} set to $|0\rangle$; the other entries represent the amplitudes for states with q_{n-1} set to $|1\rangle$. This decomposition is represented in a decision diagram structure by a node labeled q_{n-1} and two successors leading to nodes representing the sub-vectors. The sub-vectors are recursively decomposed further until vectors of size 1 (i.e., a complex number) result. This eventually represents the amplitude α_i for the complete basis state and is given by a terminal node. During this decomposition, equivalent sub-vectors are represented by the same nodes—enabling sharing and, hence, a reduction of the complexity of the representation. An example illustrates the idea.

Example 2 Consider again the quantum state from Example 1. Applying the decomposition described above yields a decision diagram as shown in Fig. 1a. The left (right) outgoing edge of each node labeled q_i points to a node representing the sub-vector with all amplitudes for the basis states with q_i set to $|0\rangle$ ($|1\rangle$). Following a path from the root to the terminal yields the respective entry. For example, following the path highlighted bold in Fig. 1a provides the amplitude for the basis state with $q_2 = |1\rangle$ (right edge), $q_1 = |1\rangle$ (right edge), and $q_0 = |0\rangle$ (left edge), i.e., $-\frac{1}{\sqrt{2}}$ which is exactly the amplitude for basis state $|110\rangle$ (seventh entry in the vector from Example 1). Since some sub-vectors are equal (e.g., $\left[\frac{1}{2}, 0\right]^T$ represented by the left node labeled q_0), sharing is possible.

However, there is more potential for sharing. In fact, many entries of the state vectors differ by a common factor only (e.g., the state vector from Example 1 has entries $\frac{1}{2}$ and $-\frac{1}{\sqrt{2}}$ which differ by the factor $-\sqrt{2}$). This is exploited in the decision diagram representation by denoting common factors of amplitudes as weights to the

---

[1]The terminology *most significant qubit* refers to a position in the basis states of a quantum system and does not signify the importance of the qubit itself.

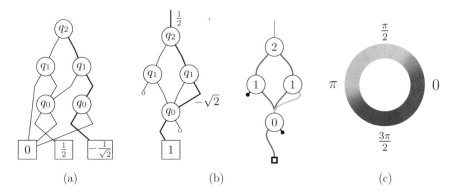

Fig. 1 Different representations of the state vector from Example 1. (**a**) Without weights. (**b**) With weights. (**c**) Graphical notation

edges of the decision diagram. Then, the value of an amplitude for a basis state is determined by not only following the path from the root to the terminal but additionally multiplying the weights of the edges along this path. Note that for a more readable notation, we use zero stubs to indicate zero vectors (i.e., vectors only containing zeroes) and omit edge weights that are equal to one. Again, an example illustrates the idea.

Example 3 Consider again the quantum state from Example 1 and the corresponding decision diagram shown in Fig. 1a. As can be seen, the sub-graphs rooting the node labeled q_0 are structurally equivalent and only differ in their terminal values. Moreover, they represent sub-vectors $\left[\frac{1}{2}, 0\right]^T$ and $\left[-\frac{1}{\sqrt{2}}, 0\right]^T$ which only differ in a common factor.

In the decision diagram shown in Fig. 1b, both sub-graphs are merged. This is possible since the corresponding value of the amplitudes is now determined not by the terminals but by the product of weights on the respective paths. As an example, consider again the path highlighted bold representing the amplitude for the basis state $|110\rangle$. Since this path includes the weights $\frac{1}{2}$, 1, $-\sqrt{2}$, and 1, an amplitude value of $\frac{1}{2} \cdot 1 \cdot (-\sqrt{2}) \cdot 1 = -\frac{1}{\sqrt{2}}$ results.

There exist various possibilities to factorize an amplitude. Hence, we apply a normalization scheme to the decision diagrams, resulting in a representation which is canonical w.r.t order of qubits [9]. The outgoing edges of a node are often normalized by dividing both weights by the weight of the leftmost edge (when $\neq 0$), and multiplying this factor to the incoming edges. However, it has been found in [11] that it is more effective to divide by the norm of the vector containing both edge weights and adjust the incoming edges accordingly. This normalizes the sum of the squared magnitudes of the outgoing edge weights to 1 and is consistent with the quantum semantics, where basis states $|0\rangle$ and $|1\rangle$ are observed after measurement with probabilities that are squared magnitudes of the respective weights. Furthermore, to ease the graphical notation we represent the complex number in polar plane as $r \cdot e^{i\alpha}$. The magnitude r of an edge weight is represented by the edge's thickness and the angle α according to the HLS color wheel [12]. The graphical notation reflects that one is most often only interested in the structure of the decision diagram instead of the exact values of edge weights. Of course, the edge weights can be put in the notation if necessary.

Example 4 Consider again the quantum state from Example 1 and the normalized decision diagram with edge weights shown in Fig. 1b. Figure 1c shows the graphical notation of this decision diagram where the line width represents the magnitude of the edge weight and the color the respecting angle when considering the polar notation of complex numbers.

In Fig. 1c, the edge to the root node (having a weight of $1/2$) is notably thinner than the other edges (with weights 1 and $-\sqrt{2}$). The with weight $-\sqrt{2}$ is slightly thicker than the edges with weight 1 and, more visible in the figure, has a different phase, i.e., $-\sqrt{2} = \sqrt{2} \cdot e^{i\pi}$, encoded in the line color.

Overall, the discussions from above lead to the following definition of decision diagrams for quantum states.

Definition 1 The decision diagram representing a 2^n-dimensional state vector is a directed acyclic multi-graph with one terminal node labeled 1 that has no successors and represents a 1-dimensional vector with the element 1. All other nodes are labeled q_i, $0 \leq i < n$ (representing a partition over qubit q_i) and have two successors. Additionally, there is an edge pointing to the root node of the decision diagram. This edge is called *root edge*. Each edge of the graph has a complex number attached as weight. An entry of the state vector is then determined by the product of all edge weights along the path from the root towards the terminal. Without loss of generality, the nodes of the decision diagram are ordered by the significance of their label, i.e., the successor of a node labeled q_i is labeled with a less significant qubit q_j. Finally, the nodes are normalized, which means that the sum of the squared magnitudes of the outgoing edge weights equals one and the common factor is propagated upwards in the decision diagram.

2.2   Representation of Matrices

While quantum states are commonly represented by vectors, quantum operations are described by matrices. These matrices are unitary (its conjugate transpose is also its inverse) and $2^n \times 2^n$-dimensional for a n-qubit system. Similar to state vectors, matrices often include redundancies, which can be exploited for a more compact representation. To this end, the decomposition scheme for state vectors is extended by a second dimension—yielding a decomposition scheme for $2^n \times 2^n$ matrices.

The entries of a unitary matrix $U = [u_{i,j}]$ indicate how much the operation U affects the mapping from a basis state $|i\rangle$ to a basis state $|j\rangle$. Considering again a quantum system with qubits $q_{n-1}, \ldots, q_1, q_0$, whereby q_{n-1} represents the most significant qubit, the matrix U is decomposed into four sub-matrices with dimension $2^{n-1} \times 2^{n-1}$: All entries in the left upper sub-matrix (right lower sub-matrix) provide the values describing the mapping from basis states $|i\rangle$ to $|j\rangle$ with both assuming $q_0 = |0\rangle$ ($q_0 = |1\rangle$). All entries in the right upper sub-matrix (left lower sub-matrix) provide the values describing the mapping from basis states $|i\rangle$ with $q_0 = |1\rangle$ to $|j\rangle$ with $q_0 = |0\rangle$ ($q_0 = |0\rangle$ to $q_0 = |1\rangle$).

This decomposition is represented in a decision diagram structure by a node labeled q_{n-1} and four successors leading to nodes representing the sub-matrices. The sub-matrices are recursively decomposed further until a 1×1 matrix (i.e., a complex number) results. This eventually represents the value $u_{i,j}$ for the corresponding mapping. Also during this decomposition, equivalent sub-matrices are represented by the same nodes and weights. As for decision diagrams representing state vectors, a corresponding normalization scheme is employed. To this end, all edges-weights are divided by the leftmost entry with the largest magnitude. Again, zero stubs are used to indicate zero matrices (i.e., matrices that contain zeros only) and edge weights equal to one are omitted. Similar to decision diagrams for quantum states, magnitude and phase of edge weights are encoded as thickness and color, respectively (see Example 4). Again, an example illustrates the idea.

Example 5 Consider the matrices of the Hadamard operation H, the identity $\mathbb{I}_2$, and their combination $U = H \otimes \mathbb{I}_2$, i.e.,

$$H = \frac{1}{\sqrt{2}} \begin{bmatrix} 1 & 1 \\ 1 & -1 \end{bmatrix} \quad \mathbb{I}_2 = \begin{bmatrix} 1 & 0 \\ 0 & 1 \end{bmatrix} \quad U = H \otimes \mathbb{I}_2 = \frac{1}{\sqrt{2}} \begin{bmatrix} 1 & 0 & 1 & 0 \\ 0 & 1 & 0 & 1 \\ 1 & 0 & -1 & 0 \\ 0 & 1 & 0 & -1 \end{bmatrix}.$$

Figure 2 shows the corresponding decision diagram representations. Following the path with dotted lines in Fig. 2c defines the entry $u_{2,0}$: a mapping from $|0\rangle$ to $|1\rangle$ for q_1 (third edge from the left) and from $|0\rangle$ to $|0\rangle$ for q_0 (first edge). Consequently the path describes the entry for a mapping from $|00\rangle$ to $|10\rangle$. Multiplying all factors on the path (including the *root edge*) yields $\frac{1}{\sqrt{2}} \cdot 1 \cdot 1 = \frac{1}{\sqrt{2}}$, which is the value of $u_{2,0}$.

Fig. 2 Representation of matrices. (**a**) H. (**b**) I_2. (**c**) $U = H \otimes I_2$

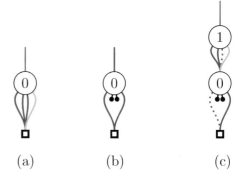

(a)   (b)   (c)

Overall, the concepts described above yield to the definition of a decision diagram representing a unitary matrix as follows.

Definition 2 The decision diagram representing a $2^n \times 2^n$-dimensional unitary matrix is a directed acyclic graph with one terminal node labeled 1 that has no successors and represents a 1×1 matrix with the element 1. All other nodes are labeled q_i, $0 \leq i < n$ (representing a partition over qubit q_i) and have four successors. Additionally, there is an edge pointing to the root node of the decision diagram. This edge is called *root edge*. Each edge of the graph has attached a complex number as weight. An entry of the unitary matrix is then determined by the product of all edge weights along the path from the root towards the terminal. Without loss of generality, the nodes of the decision diagram are ordered by the significance of their label, i.e., the successor of a node labeled q_i are labeled with a less significant qubits q_j. Finally, the nodes are normalized, which means that all edges-weights are divided by the leftmost entry with the largest magnitude. The common factor is propagated upwards in the decision diagram.

A performance-oriented implementation handling decision diagrams and operations as described in this section is freely available under the MIT license at https://github.com/cda-tum/dd_package. Furthermore, to give a better intuition and make decision diagrams for quantum computing more accessible, an installation-free web-tool that visualizes decision diagrams for state vectors as well as matrices is available at https://www.cda.cit.tum.de/app/ddvis/.

Given the decision diagrams as described in this section, the following sections showcase the applicability in different areas of design and verification work. More precisely, we cover the simulation of quantum circuits without noise in Sect. 3 and with noise in Sect. 4 as well as verification of quantum circuits in Sect. 5.

3 Simulation of Quantum Circuits

Despite physical quantum computers being available in the cloud nowadays, the simulation of quantum circuits on non-quantum hardware remains paramount for the development and design of future quantum computing applications. Additionally, simulations on non-quantum hardware provide insights into the inner workings of a quantum system that are fundamentally hidden in physical quantum computers. This enables designers to analyze quantum algorithms or verify the output of physical quantum computers. To this end, simulating a quantum circuit entails the successive application of all individual gates of the circuit to the initial state of a quantum system in order to obtain the final state. The final state is measured to obtain the result in the computational bases. While straight-forward in principle, this quickly amounts to a hard task due to the required memory on non-quantum hardware and the subsequently difficult manipulation of 2^n complex amplitudes for an n-qubit system.

Decision diagrams, as described in Sect. 2, provide a promising technique that aims at compactly representing the 2^n complex amplitudes of a quantum system and the corresponding operations applied to it. Having the ability to compactly represent state vectors and unitary matrices, all that is left is to provide corresponding methods to form the Kronecker product, multiply vectors with matrices, as well as measure the quantum system. Since the introduced decision diagrams closely relate to vectors and matrices, we can implement the required operations by slight adaptations only.

3.1 Kronecker Product

The Kronecker product enables composition of multiple matrices to attain the suitable size $2^n \times 2^n$ matrix to be applied to an n-qubit system. Given two matrices A and B, the Kronecker product is defined as in Eq. (1).

$$A \otimes B = \begin{bmatrix} a_{0,0} \cdot B & \cdots & a_{0,2^k-1} \cdot B \\ \vdots & \ddots & \vdots \\ a_{2^k-1,0} \cdot B & \cdots & a_{2^k-1,2^k-1} \cdot B \end{bmatrix} \tag{1}$$

In other words, the Kronecker product replaces each element $a_{i,j}$ of A by $a_{i,j} \cdot B$. While this constitutes a computationally expensive task using straight-forward realizations by means of array-based implementations of A and B, it is very cheap to form the Kronecker product of two matrices given as decision diagrams.

Since $a_{i,j}$ is given as product of the edge weights from A's root node to the terminal and we can easily determine $a_{i,j} \cdot B$ by adjusting the weight of the edge pointing to B's root node. All that has to be done to determine $A \otimes B$ is replacing A's terminal with the root node of B. Additionally, the weight of A's root edge has to be multiplied by the weight of B's root edge.

Example 6 Recall the matrices considered in Fig. 2c. The Kronecker product $U = H \otimes \mathbb{I}_2$ can efficiently be constructed by taking the decision diagram representation of H and replacing its terminal node with the root node of the decision diagram representing $\mathbb{I}_2$. Since the root edge of $\mathbb{I}_2$ has weight 1, the value of the root node of U is equal to the weight of A's root edge. This is illustrated in Fig. 3.

Fig. 3 Creation of $H \otimes \mathbb{I}_2$ using decision diagrams

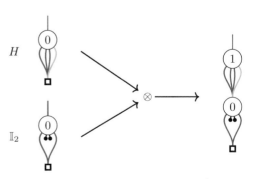

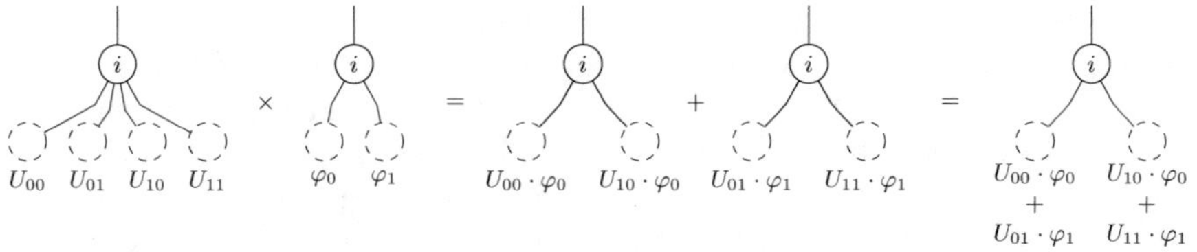

Fig. 4 Recursive structure of multiplication and addition using decision diagrams

3.2 Adding and Multiplying Unitary Matrices

The multiplication of a unitary matrix U and a state vector $|\varphi\rangle$ can be broken down into sub-computations according to Eq. (2).

$$\begin{bmatrix} U_{00} & U_{01} \\ U_{10} & U_{11} \end{bmatrix} \cdot \begin{bmatrix} \varphi_0 \\ \varphi_1 \end{bmatrix} = \begin{bmatrix} (U_{00}\varphi_0 + U_{01}\varphi_1) \\ (U_{10}\varphi_0 + U_{11}\varphi_1) \end{bmatrix} \tag{2}$$

For decision diagrams, recursively determining the four sub-products $U_{00} \cdot \varphi_0$, $U_{01} \cdot \varphi_1$, $U_{10} \cdot \varphi_0$, and $U_{11} \cdot \varphi_1$ realizes the multiplication. The decompositions of multiplication and addition are recursively applied until 1×1 matrices or 1-dimensional vectors result. Since these represent just complex numbers, their multiplication and addition is well defined.

As shown in the middle of Fig. 4, these sub-products are then combined with a decision diagram node to two intermediate state vectors. Finally, these intermediate state vectors have to be added. This addition is recursively decomposed similarly, namely as in Eq. (3).

$$\psi + \phi = \begin{bmatrix} \psi_0 \\ \psi_1 \end{bmatrix} + \begin{bmatrix} \phi_0 \\ \phi_1 \end{bmatrix} = \begin{bmatrix} \psi_0 + \phi_0 \\ \psi_1 + \phi_1 \end{bmatrix} \tag{3}$$

The recursively determined sub-sums $\psi_0 + \phi_0$ and $\psi_1 + \phi_1$ are composed by a decision diagram node as shown on the right-hand side of Fig. 4.

Moreover, decomposition into sub-products and sub-sums does not change the decision diagram structure. Hence, the complexity of them remains bounded by the number of nodes of the original representations. Furthermore, redundancies can again be exploited by caching sub-products and sub-sums.

3.3 Measurement

Measurement can efficiently be conducted on the decision diagram structure. Without loss of generality, consider that the most significant qubit (which is represented by the root node of the corresponding decision diagram) of the state vector should be measured. This can be accomplished by applying a SWAP operation or by re-arranging the nodes and edges of the decision diagram. Then, the probability of choosing either the left or right edge is given by the *upstream* probability of the successor nodes weighted by the corresponding edge weights [13]. Depending on the used normalization scheme, this calculation may be simplified [14]. An example illustrates the idea.

Example 7 Consider again the quantum state discussed in Example 1 and its corresponding decision diagram shown in Fig. 1b. Then, the probabilities for measuring $q_2 = |0\rangle$ and $q_2 = |1\rangle$ are $|\frac{1}{2}|^2 \cdot |1|^2 \cdot 1 = \frac{1}{4}$ and $|\frac{1}{2}|^2 \cdot |1|^2 \cdot 3 = \frac{3}{4}$, respectively.

Having the probabilities for collapsing the most significant qubit to basis state $|0\rangle$ and $|1\rangle$ allows to sample its new value. If we obtain basis state $|0\rangle$ ($|1\rangle$), the amplitudes for all basis states with $q_{n-1} = |1\rangle$ ($q_{n-1} = |0\rangle$) are set to zero. In the decision diagram, the collapse is performed by multiplying with the M_0 (M_1) non-unitary matrix from Eq. (4).

$$M_0 = \begin{bmatrix} 1 & 0 \\ 0 & 0 \end{bmatrix} \quad M_1 = \begin{bmatrix} 0 & 0 \\ 0 & 1 \end{bmatrix} \tag{4}$$

Afterwards the decision diagram is renormalized and the probability of the con-sidered qubit being measurement again in the same basis state in subsequent measurements is 1.

Example 7 (continued) Assume we measure basis state $|1\rangle$ for qubit q_2. Figure 5a shows the resulting decision diagram. To fulfil the normalization constraint, we renormalize the decision diagram—eventually resulting in the decision diagram shown in Fig. 5b.

Fig. 5 Measurement of qubit q_2. (**a**) Measure $q_2 = |1\rangle$. (**b**) Normalize amplitudes

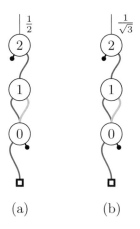

(a) (b)

Measuring all qubits can be conducted in a similar fashion. In fact, we repeat the procedure discussed above sequentially for all qubits $q_{n-1}, q_{n-2}, \ldots, q_0$. Assume that qubit q_i shall be measured, and that all qubits q_j where $j > i$ are already measured. Then, there exists only one node labeled q_i, which is the root node of the sub-vector to be measured.

Overall, this allows for an more efficient quantum circuit simulation in many cases. An implementation of this method is available at https://github.com/cda-tum/ddsim and via the corresponding Python package `mqt.ddsim`. In addition, https://www.cda.cit.tum.de/app/ddvis/ provides an installation-free visualization scheme of the procedure.

4   Noise-Aware Simulation of Quantum Circuits

The methods reviewed above allow for an efficient simulation of *perfect*, i.e., error-free, quantum circuits. While already an important step towards proper design and evaluation of certain applications, physical quantum computers do not work perfectly and are affected by noise effects, which cause errors during quantum computations. Considering those errors during simulation enables a more accurate and realistic evaluation of the, respectively, considered quantum circuits. At the same time, considering errors introduces new challenges for the already exponentially hard problem of quantum circuit simulation. In this section, we review how to conduct noisy simulation with decision diagrams. To this end, we first review typical noise effects in quantum computing, i.e., recap frequently occurring errors, and, afterwards, discuss two complementary solutions for noise-aware simulation based on decision diagrams (originally proposed in [15, 16]).

4.1   Errors in Quantum Computations

During quantum computations, a large variety of errors may occur and affect the output of the corresponding executions. Most prominently, two types of errors are distinguished [17]:

Gate errors: Any errors that may alter the originally intended functionality of an operation or may lead to an operation not being executed at all.

Decoherence errors: Any errors caused by the effect that qubits can only hold information for a limited amount of time.

Gate errors heavily depend on the underlying quantum computer technology and even on the qubits to which the respective operations are applied. The effect is that the operation either is not executed at all or that a different operation is employed. Often, they are approximated using depolarization errors [18, 19] and, hence, defined by altering the qubit to a completely random state [20]. More detailed

descriptions of the respective effects are additionally often provided by the vendors
of the respective quantum computer, e.g., in case of IBM at [21].

Decoherence Errors occur due to the fragile nature of quantum systems. Because
of this, qubits can only hold information for a limited time and, hence, qubits in
a high-energy state ($|1\rangle$) tend to relax towards a low energy state ($|0\rangle$) (i.e., after
a certain amount of time, qubits eventually decay towards $|0\rangle$). Moreover, when a
qubit interacts with the environment, further errors (such as phase-flip errors) might
occur.

Example 8 Consider a 2-qubit system which is in state $\left|\psi'\right\rangle = \frac{1}{\sqrt{2}}(|00\rangle + |11\rangle)$
and assume that a gate error might affect this state with probability p. Then, with
probability $1 - p$, nothing happens (the state remains unchanged) while, with
probability p, a certain error effect is imposed. Both scenarios can be captured by
either employing an I-operation or an operation describing the error effect (e.g., a
polarization using X, Y, or Z or a completely random effect), respectively.

In a similar fashion, consider the same quantum state but assume a decoherence
error at the second qubit (more precisely, a dampening error which makes the second
qubit decay to $|0\rangle$) with a probability $p = 0.3$. Then, a measurement of this state
would not lead to $|00\rangle$ or $|11\rangle$ with equal probabilities anymore (as in the error-free
case), but to $|00\rangle$ in 50% of the cases, $|10\rangle$ in 15% of the cases, and $|11\rangle$ in 35% of
the cases. That is, the probability that the second qubit decays to $|0\rangle$ is substantially
larger due to the decoherence.

Overall, errors effects can be seen as (unwanted) operations employed on a
quantum system. Accordingly, they could in principle be simulated like any other
quantum operation—using, e.g., the methods described before in Sect. 3. The main
challenge, however, is that whether an error effect happens or not depends on certain
probabilities. These need to be captured during the simulation. Existing solutions
doing that have been proposed, e.g., in [19, 22–29].

4.2 Simulation Methods Using Decision Diagrams

In this section, we review how decision diagrams may help in providing a solution
for noise-aware quantum circuit simulation. The first solution thereby relies on a
stochastic approach which employs the main concepts reviewed in Sect. 3, while
the second solution aims for a deterministic consideration of noise—requiring a
more complex representation of quantum states and operations.

More precisely, the first solution (stochastic simulation as proposed in [16]) is
based on the vanilla decision diagram-based circuit simulator described in Sect. 3.
Then, whenever the considered quantum computer might make an error during
its simulated operation, it either mimics the effect of the error by additionally
employing an error operation (with corresponding probability p) or leaves the state

untouched (with probability $1 - p$). Consequently, a *single* output state is sampled from the whole spectrum of possible output states by such a run.

By iteratively sampling sufficiently many output states (using, e.g., stochastic Monte-Carlo approximation) a rather accurate approximation of the quantum circuit's behavior under the influence of noise effects can be obtained. The benefit of this approach is that it does not substantially increase the complexity of individual simulation runs when compared to the error-free circuit simulation. Furthermore, individual simulations are independent and, hence, can be executed in parallel. However, this approach remains stochastic, i.e., it cannot guarantee the best possible accuracy (although evaluations summarized in [30] show that a sufficient accuracy can be achieved for practically relevant use cases).

If an exact consideration of noise is desired, a more elaborate solution is required which describes all noise effects in a deterministic fashion. To this end, the representation of quantum states and quantum operations in terms of vectors and matrices (as used thus far) is not sufficient anymore. More precisely, a description is needed which incorporates all possible states a quantum system may reside in (including the original state but also states resulting from any noise effects with certain probabilities).

This is accomplished by extending the state vector representation to *density matrices* (also known as *density operators*) [31]. More precisely, let $|\phi\rangle$ be a complex vector representing the state of a quantum system. Then, the corresponding *density matrix* is defined as $\rho = |\phi\rangle \langle\phi|$ with $\langle\phi| := |\phi\rangle^\dagger$.

Example 9 Consider again the quantum state $|\psi'\rangle = \frac{1}{\sqrt{2}}(|00\rangle + |11\rangle)$ from Example 8. The corresponding *density matrix* ρ is given by

$$\begin{bmatrix} \frac{1}{\sqrt{2}} \\ 0 \\ 0 \\ \frac{1}{\sqrt{2}} \end{bmatrix} \cdot \begin{bmatrix} \frac{1}{\sqrt{2}} & 0 & 0 & \frac{1}{\sqrt{2}} \end{bmatrix} = \begin{bmatrix} \frac{1}{2} & 0 & 0 & \frac{1}{2} \\ 0 & 0 & 0 & 0 \\ 0 & 0 & 0 & 0 \\ \frac{1}{2} & 0 & 0 & \frac{1}{2} \end{bmatrix}. \tag{5}$$

This representation properly describes the quantum state while, additionally, allowing to store information about the noise effects on the state. For example, the diagonal entries encode the probabilities for measuring $|00\rangle$, $|01\rangle$, $|10\rangle$, and $|11\rangle$, respectively, which is in-line with the probabilities obtained from the state vector representation ($|\frac{1}{\sqrt{2}}|^2 = \frac{1}{2}$).

Based on this representation, various error effects can now be applied by a tuple $(E_0, E_1, \ldots, E_m)$ of *Kraus matrices* satisfying the condition

$$\sum_{i=0}^{m} E_i^\dagger E_i = \mathbb{I}. \tag{6}$$

For example, a decoherence error which makes a qubit decay to $|0\rangle$ can be represented by [20]

$$(E_0, E_1) \text{ with } E_0 = \begin{bmatrix} 1 & 0 \\ 0 & \sqrt{1-p} \end{bmatrix} \text{ and } E_1 = \begin{bmatrix} 0 & \sqrt{p} \\ 0 & 0 \end{bmatrix}, \tag{7}$$

where the variable p represents the probability of the error occurring. Other noise effects can be described in a similar fashion. Applying these error descriptions to a quantum system given by the density matrix ρ yields the density matrix [20]:

$$\rho' = \sum_{i=0}^{m} E_i \rho E_i^{\dagger}. \tag{8}$$

This formalism allows to deterministically capture all noise effects as illustrated in the following example:

Example 10 Consider again the quantum state from above and the decoherence error from Example 8 making the second qubit decay to $|0\rangle$) with probability $p = 0.3$. Then, the error's effect can be calculated deterministically by constructing the correctly sized matrices with the Kronecker product and summing them as in Eq. (9).

$$\underbrace{\begin{bmatrix} 0.5 & 0 & 0 & 0.418 \\ 0 & 0 & 0 & 0 \\ 0 & 0 & 0 & 0 \\ 0.418 & 0 & 0 & 0.35 \end{bmatrix}}_{E_0 \rho E_0^{\dagger}} + \underbrace{\begin{bmatrix} 0 & 0 & 0 & 0 \\ 0 & 0 & 0 & 0 \\ 0 & 0 & 0.15 & 0 \\ 0 & 0 & 0 & 0 \end{bmatrix}}_{E_1 \rho E_1^{\dagger}} = \underbrace{\begin{bmatrix} 0.5 & 0 & 0 & 0.418 \\ 0 & 0 & 0 & 0 \\ 0 & 0 & 0.15 & 0 \\ 0.418 & 0 & 0 & 0.35 \end{bmatrix}}_{\rho'} \tag{9}$$

Again the diagonal encodes the probabilities for measuring $|00\rangle$, $|01\rangle$, $|10\rangle$, and $|11\rangle$, which are in-line with the values covered before in Example 8.

The resulting representations are substantially larger than the vectors and matrices needed for error-free quantum circuit simulation. For example, rather than vectors of size 2^n, matrices of size $2^n \times 2^n$ are needed to represent an n-qubit quantum state. However, density matrices can be represented in terms of decision diagrams as well. In fact, employing the same decomposition scheme for matrices as reviewed in Sect. 3 yields corresponding decision diagrams for density matrices.

Example 11 Consider again the quantum state from Example 9 in both the vector as well as density matrix representation. The corresponding decision diagram representations are provided in Fig. 6a and b, respectively. The decision diagram resulting after applying the error affect (as considered in Example 10) is shown in Fig. 6c.

Obviously, the decision diagram representations of the density matrices (providing a deterministic representation of all employed error effects) are larger than the

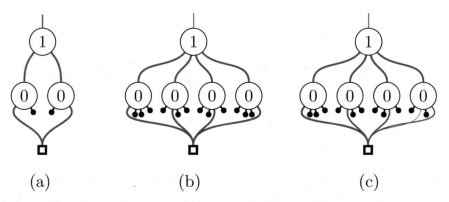

Fig. 6 Decision diagram representation of states. (**a**) Decision diagram of a state vector. (**b**) Decision diagram of a density matrix. (**c**) State after applying T1 error

original state representation. After all, a substantially larger amount of information needs to be stored. Nevertheless, the examples show that, also in these cases, decision diagrams may offer a more compact representation than offered by a direct representation in terms of a $2^n \times 2^n$-matrix. After all, this helps in improving deterministic, noise-aware quantum circuit simulation.

Overall, this section showed that decision diagrams can be employed for noise-aware quantum circuit simulation—both, stochastically as well as deterministically. Further details and evaluations on the respective methods are available in [15, 16]. An implementation of the stochastic method is available at https://github.com/cda-tum/ddsim.

5   Verification of Quantum Circuits

As a final example for the utilization of decision diagrams in design/software for quantum computing, we consider the task of verification—more precisely equivalence checking. Here, the question is whether two quantum circuits G and G' realize the same functionality. This is motivated by the design flow in which a given circuit is decomposed, mapped, and optimized [32–42]. During all these steps, it has to be ensured that the functionality of the correspondingly resulting circuit descriptions does not change. In the following, we first give an explicit description of this problem and, then, describe two complementary approaches for tackling it using decision diagrams.

5.1   The Quantum Circuit Equivalence Checking Problem

Equivalence checking in the domain of quantum computing—as we consider it in this work—is about proving that two quantum circuits G and G' are functionally equivalent (i.e., realize the same function), or to show the non-equivalence of

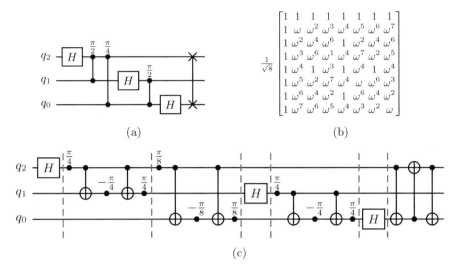

Fig. 7 The QFT, its functionality, and an alternative realization. (**a**) QFT Circuit G. (**b**) Functionality U ($\omega = (1 + i)/\sqrt{2}$). (**c**) Alternative realization G'

these circuits by means of a counterexample. To this end, consider two quantum circuits $G = g_0 \ldots g_{m-1}$ and $G' = g'_0 \ldots g'_{m'-1}$ operating on n qubits. Then, the functionality of each circuit can be uniquely described by the respective system matrices $U = U_{m-1} \cdots U_0$ and $U' = U'_{m'-1} \cdots U'_0$, where the matrices $U_i^{(')}$ describe the functionality of the i-th gate of the respective circuit (with $0 \leq i < m^{(')}$). Consequently, deciding the equivalence of both computations amounts to comparing the system matrices U and U'. More precisely, U and U' are considered equivalent, if they at most differ by a global phase factor (which is fundamentally unobservable [43]), i.e., $U = e^{i\alpha} U'$ with $\alpha \in [0, 2\pi)$.

Example 12 Consider the circuit G of the three qubit Quantum Fourier Transform shown in Fig. 7a. Its corresponding functionality is described by the densely populated 8×8 matrix U shown in Fig. 7b [43]. Additionally, Fig. 7c shows an alternative realization G' of the functionality of G. Since both circuits exhibit the same system matrix U, they are considered equivalent.

Unfortunately, the whole functionality U (and similarly U') is not readily available for performing this comparison but has to be constructed from the individual gate descriptions g_i—requiring the subsequent matrix-matrix multiplications $U^{(0)} = U_0$, $U^{(j)} = U_j \cdot U^{(j-1)}$ for $j = 1, \ldots, m - 1$ to construct the whole system matrix $U = U^{(m-1)}$. While conceptually simple (as the matrix-vector multiplication for simulation discussed in Sect. 3), this quickly constitutes an extremely complex task due to the exponential size of the involved matrices with respect to the number of qubits. In fact, equivalence checking of quantum circuits has been shown to be QMA-complete [44].

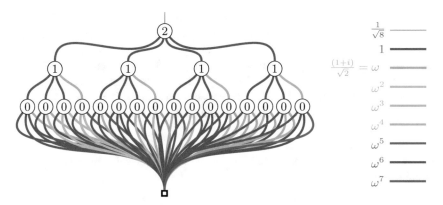

Fig. 8 Decision diagram for the functionality U of G shown in Fig. 7

Due to their potential for compactly representing and efficiently manipulating the functionality of a quantum circuit, decision diagrams are a perfect fit for this task. However, merely using decision diagrams to construct a representation of both circuits' functionality and comparing them still has significant shortcomings. In fact, representing the entire functionality of a quantum circuit still might be exponential in the worst case. However, this can be addressed by additionally exploiting the reversibility of quantum circuits.

Example 13 Consider the functionality U of the QFT circuit G from Fig. 7. Its corresponding decision diagram (shown in Fig. 8 with color legend to the right-hand side) is as densely populated as the matrix it represents since no redundancy can be exploited, i.e., each node's successors point to unique child nodes.

5.2 Exploiting Reversibility

Most classical logic operations are not reversible (e.g., neither $x \wedge y = 0$ nor $x \vee y = 1$ allows one to determine the values of x and y). As there is no bijective mapping between input and output states, in general, the concept of the *inverse* of a classical operation (or a sequence thereof) is not meaningful. In contrast, all quantum operations are inherently *reversible*. Consider an operation g described by the unitary matrix U. Then, its inverse U^{-1} is efficiently calculated as the conjugate-transpose $U^{\dagger}$. Given a sequence of m operations $g_0, \ldots, g_{m-1}$ with associated matrices $U_0, \ldots, U_{m-1}$, the inverse of the corresponding system matrix $U = U_{m-1} \cdots U_0$ is derived by reversing the operations' order and inverting each individual operation, i.e., $U^{-1} = U^{\dagger} = U_0^{\dagger} \cdots U_{m-1}^{\dagger}$.

This characteristics can be exploited to improve the performance of the verification approach presented above. To this end, consider two quantum circuits G and G'. In case both circuits are functionally equivalent, this allows for the conclusion that concatenating one circuit with the inverse of the other realizes the identity function $\mathbb{I}$, i.e., $G'^{-1} \cdot G \equiv \mathbb{I}$. This offers significant potential since the identity constitutes the best case for decision diagrams (the identity can be represented by a decision diagram of linear size). Unfortunately, creating such a concatenation in a naive fashion, e.g., by computing $U \cdot U'^\dagger$ hardly yields any advantage because, even if the final decision diagram would be as compact as possible, the full (and potentially exponential) decision diagram of at least one of the circuits would be generated as an intermediate result.

Instead, the full potential of this observation is utilized if the associativity of the respective multiplications is fully exploited. More precisely, given two quantum circuits G and G', it holds that

$$G'^{-1} \cdot G = (g'^{-1}_{m'-1} \cdots g'^{-1}_0) \cdot (g_0 \cdots g_{m-1})$$

$$\equiv (U_{m-1} \cdots U_0) \cdot (U'^\dagger_0 \cdots U'^\dagger_{m'-1})$$

$$= U_{m-1} \cdots U_0 \cdot \mathbb{I} \cdot U'^\dagger_0 \cdots U'^\dagger_{m'-1}$$

$$=: G \rightarrow \mathbb{I} \leftarrow G'.$$

Here, $G \rightarrow \mathbb{I} \leftarrow G'$ symbolizes that, starting from the identity $\mathbb{I}$, either gates from G can be "applied from the left" or (inverted) gates from G' can be "applied from the right." If the respective gates of G and G' are applied in a fashion frequently yielding the identity, the entire equivalence checking process can be conducted on rather small (intermediate) decision diagrams. This is illustrated by the following example.

Example 14 Consider again the two circuits G and G' from Example 12 and assume that, starting with a decision diagram representing the identity, for every gate applied from G all gates from G' until the next red barrier shown in Fig. 7f are applied. Applying the gates from G and G' in such a particular order "from the left" and "from the right," respectively, yields situations where the impact of a gate from circuit G (increasing the size of the decision diagram) is reverted by multiplications with inverted gates from G' (decreasing the size of the decision diagram back to the representation of the identity function). This way, the equivalence check can be conducted on much smaller intermediate representations and, hence, much more efficiently.

Moreover, even if the considered circuits G and G' are *not* functionally equivalent (and, hence, identity is not achieved), the observations from above still promise improvements compared to creating the complete decision diagrams for G and G'. This is, because in this case, the result of $G \rightarrow \mathbb{I} \leftarrow G'$ inherently provides an efficient representation of the circuit's difference that allows one to obtain counterexamples almost "for free" (while those have to be explicitly generated using additional inversion and multiplication operations otherwise).

Overall, following those ideas, equivalence checking of two quantum circuits can be conducted very efficiently on rather compact decision diagrams, as shown in [45]. But determining when to apply gates from G and when to apply (inverted) gates from G' is not at all obvious. Designing dedicated strategies for specific applications is a topic of ongoing research. As an example, a dedicated strategy for verifying the results of compilation flows can be derived by exploiting knowledge about the compilation flow itself [46]. An implementation of this method is available at https://github.com/cda-tum/qcec and via the corresponding Python package mqt.qcec. In addition, https://www.cda.cit.tum.de/app/ddvis/ provides an installation-free visualization scheme of the procedure that also can be used to try out different gate-application schemes.

5.3 The Power of Simulation

The second characteristic we are exploiting rests on the observation that simulation is much more powerful for equivalence checking of quantum circuits than for equivalence checking of classical circuits. More precisely, in the classical realm, it is certainly possible to simulate two circuits with random inputs to obtain counterexamples in case they are not equivalent. However, this often does not yield the desired result. In fact, due to masking effects and the inevitable information loss introduced by many classical gates, the chance of detecting differences in the circuits within a few arbitrary simulations is greatly reduced (e.g., $x \wedge 0$ masks any difference that potentially occurs during the calculation of x). Consequently, sophisticated schemes for constraint-based stimuli generation [47–50], fuzzing [51, 52], etc. are employed in order to verify classical circuits.

In quantum computing, the inherent reversibility of quantum operations dramatically reduces these effects and frequently yields situations where even small differences remain unmasked and affect entire system matrices—showing the power of random simulations for checking the equivalence of quantum circuits. Because of that, it is in general not necessary to compare the *entire* system matrices—in particular when two circuits are *not* equivalent and, hence, their system matrices differ from each other.

Given two unitary matrices U and U', we define their *difference* D as the unitary matrix $D = U^\dagger U'$ and it holds that $U \cdot D = U'$. In case both matrices are identical (i.e., the circuits are equivalent), it directly follows that $D = \mathbb{I}$. One characteristic of the identity function $\mathbb{I}$ resulting in this case is that all diagonal entries are equal to one, i.e., $\langle i | U^\dagger U' | i \rangle = 1$ for $i \in \{0, \ldots, 2^n - 1\}$, where $|i\rangle$ denotes the ith computational basis state. More generally—in case of a potential relative/global phase difference between G and G'—all diagonal elements have modulus one, i.e., $| \langle i | U^\dagger U' | i \rangle |^2 = 1$. This expression can further be rewritten to

$$1 = | \langle i | U^\dagger U' | i \rangle |^2 = |(U | i \rangle)^\dagger (U' | i \rangle)|^2 = | |u_i\rangle^\dagger |u_i'\rangle |^2$$
$$= | \langle u_i | u_i' \rangle |^2,$$

where $|u_i\rangle$ and $|u_i'\rangle$ denote the ith column of U and U', respectively. This essentially resembles the simulation of both circuits with the initial state $|i\rangle$ and, afterwards, calculating the fidelity $\mathcal{F}$ between the resulting states $|u_i\rangle$ and $|u_i'\rangle$. Hence, if only one simulation yields $\mathcal{F}_i := \mathcal{F}(|u_i\rangle, |u_i'\rangle) \not\approx 1$, then $|i\rangle$ proves the non-equivalence of G and G'.

This constitutes an exponentially easier task than constructing the entire system matrices U and U'—although the complexity of simulation still remains exponential with respect to the number of qubits (for which the decision diagram-based solution reviewed above provides an efficient solution in many cases). Regarding the complexity, creating the entire system matrices corresponds to simulating the respective circuit with all 2^n different computational basis states. All this, of course, does not guarantee that any difference is indeed detected by just simulating a limited number of arbitrary computational basis states $|i\rangle$. This brings up the following question: How significantly do the matrices U and U' differ from each other in case of non-equivalence, i.e., how many computational basis states $|i\rangle$ yield $\mathcal{F}_i \not\approx 1$ for a given difference matrix D. Since the difference D of both matrices is unitary itself, it may as well be interpreted as a quantum circuit G_D. In the following, we assume that each gate of G_D either represents a single-qubit or a multi-controlled operation.[2]

Example 15 Assume that G_D only consists of one (non-trivial) single-qubit operation defined by the matrix U_s applied to the first of n qubits. Then, the system matrix D is given by $diag(U_s, \ldots, U_s)$. The process of going from U to U', i.e., calculating $U \cdot D$, impacts *all* columns of U. Thus, an error is detected by a *single* simulation with *any* computational basis state.

Among all quantum operations, single-qubit operations possess a system matrix least similar to the identity matrix due to the tensor product structure of their corresponding system matrix.

Example 16 In contrast to Example 15, assume that G_D only consists of one operation U_s targeted at the first qubit and controlled by the remaining $n - 1$ qubits. Then, the corresponding system matrix is given by $diag(\mathbb{I}_2, \ldots, \mathbb{I}_2, U_s)$. In this case, applying D to U only affects the last two columns of U. As a consequence, a maximum of two columns (out of 2^n) may serve as counterexamples—the worst case scenario.

These basic examples cover the extreme cases when it comes to the difference of two unitary matrices. In case G_D exhibits no such simple structure, the analysis is more involved, e.g., generally quantum operations with $c \in \{0, \ldots, n - 1\}$ controls will exhibit a difference in 2^{n-c} columns. Furthermore, given two operations showing a certain number of differences, the matrix product of these operations

---

[2]This does not limit the applicability of the following findings, since arbitrary single-qubit operations combined with CNOT form a universal gate-set [43].

in most cases (except when cancellations occur) differs in as many columns as the maximum of both operands.

The gate-set provided by (current) quantum computers typically includes only (certain) single-qubit gates and a specific two qubit gate, such as the *CNOT* gate. Thus, multi-controlled quantum operations usually only arise at the most abstract algorithmic description of a quantum circuit and are then *decomposed* into elementary operations from the device's gate-set before the circuit is mapped to the target architecture. As a consequence, errors occurring during the design flow will typically consist of (1) single-qubit errors, e.g., offsets in the rotation angle, or (2) errors related to the application of *CNOT* or *SWAP* gates. In both cases, non-equivalence can be efficiently concluded by a limited number of simulations with arbitrary computational basis states. If a counterexample was not obtained after a few simulations, this yields a highly probable estimate of the circuit's equivalence— in contrast to the classical realm, where this generally does not allow for any conclusion. Further details and evaluations on the respective methods are available in [45, 53].

6 Conclusions

The power of quantum computing comes with new computing primitives and the need for suitable design automation methods. In this work, we reviewed decision diagrams for quantum computing as well as their application in quantum circuit simulation (with and without noise) as well as the verification of quantum circuits. Decision diagrams offer a complementary approach for tackling the complexity of these tasks with a potential impact comparable to their conventional counterparts. With this work, we want to encourage their usage in the quantum (design automation) community. Implementations of the methods presented here are available at the corresponding GitHub repositories mentioned above.

Acknowledgments We sincerely thank all co-authors and collaborators who work(ed) with us in this exciting area. Special thanks go to Alwin Zulehner and Thomas Grurl.

This work received funding from the European Research Council (ERC) under the European Union's Horizon 2020 research and innovation program (grant agreement No. 101001318), was part of the Munich Quantum Valley, which is supported by the Bavarian state government with funds from the Hightech Agenda Bayern Plus, and has been supported by the BMWK on the basis of a decision by the German Bundestag through project QuaST, as well as by the BMK, BMDW, and the State of Upper Austria in the frame of the COMET program (managed by the FFG).

References

1. J. Preskill, Quantum computing in the NISQ era and beyond. Quantum **2**, 79 (2018)
2. R.E. Bryant, Graph-based algorithms for Boolean function manipulation. IEEE Trans. Comput. **C-35**(8), 677–691 (1986)
3. R.E. Bryant, Y.-A. Chen, Verification of arithmetic circuits using binary moment diagrams. Softw. Tools Tech. Transfer **3**(2), 137–155 (2001)

4. S. Minato, Zero-suppressed BDDs for set manipulation in combinatorial problems, in *Design Automation Conf.*, 1993, pp. 272–277
5. T. van Dijk, R. Wille, R. Meolic, Tagged BDDs: Combining reduction rules from different decision diagram types, in *Int'l Conf. on Formal Methods in CAD*, 2017, pp. 108–115
6. S.-A. Wang, C.-Y. Lu, I.-M. Tsai, S.-Y. Kuo, An XQDD-based verification method for quantum circuits, in *IEICE Trans. Fundamentals*, 2008, pp. 584–594
7. A. Abdollahi, M. Pedram, Analysis and synthesis of quantum circuits by using quantum decision diagrams, in *Design, Automation and Test in Europe*, 2006
8. G.F. Viamontes, I.L. Markov, J.P Hayes, High-performance QuIDD-Based simulation of quantum circuits, in *Design, Automation and Test in Europe*, 2004
9. D. Miller, M. Thornton, QMDD: A decision diagram structure for reversible and quantum circuits, in *Int'l Symp. on Multi-Valued Logic*, 2006
10. P. Niemann, R. Wille, D.M. Miller, M.A. Thornton, R. Drechsler, QMDDs: Efficient quantum function representation and manipulation. IEEE Trans. CAD Integr. Circuits Syst. **35**(1), 86–99 (2016)
11. S. Hillmich, I.L. Markov, R. Wille, Just like the real thing: Fast weak simulation of quantum computation, in *Design Automation Conf.,* 2020
12. R. Wille, L. Burgholzer, M. Artner, Visualizing decision diagrams for quantum computing, in *Design, Automation and Test in Europe*, 2021
13. A. Zulehner, R. Wille, Advanced simulation of quantum computations. IEEE Trans. CAD Integr. Circuits Syst. **38**(5), 848–859 (2019)
14. S. Hillmich, I.L. Markov, R. Wille, Just like the real thing: Fast weak simulation of quantum computation, in *Design Automation Conf.*, 2020
15. T. Grurl, J. Fuß, R. Wille, Considering decoherence errors in the simulation of quantum circuits using decision diagrams, in *Int'l Conf. on CAD*, 2020
16. T. Grurl, R. Kueng, J. Fuß, R. Wille, Stochastic quantum circuit simulation using decision diagrams, in *Design, Automation and Test in Europe*, 2021
17. S.S. Tannu, M.K. Qureshi, Not all qubits are created equal: A case for variability-aware policies for NISQ-era quantum computers, in *Int'l Conf. on Architectural Support for Programming Languages and Operating Systems*, 2019, pp. 987–999
18. N. Khammassi, I. Ashraf, X. Fu, C.G. Almudéver, K. Bertels, QX: A high-performance quantum computer simulation platform, in *Design, Automation and Test in Europe*, ed. by D. Atienza, G.D. Natale, 2017, pp. 464–469
19. H. Abraham, et al. *Qiskit: An Open-Source Framework for Quantum Computing* (2019)
20. M.A. Nielsen, I.L. Chuang, *Quantum Computation and Quantum Information (10th Anniversary edition)* (Cambridge University Press, 2016)
21. J. Gambetta, S. Sheldon, *Cramming More Power into a Quantum Device* https://www.ibm.com/blogs/research/2019/03/power-quantum-device/, Accessed: 2021-04-08, 2019
22. S.E. Atos, *Quantum Learning Machine*, atos.net/en/products/quantum-learning-machine. Accessed: 2021-04-08, 2016
23. N. Khammassi, I. Ashraf, X. Fu, C. Almudever, K. Bertels, QX: A high-performance quantum computer simulation platform, in *Design, Automation and Test in Europe*, 2017
24. D. Wecker, K.M. Svore, LIQUi|>: A software design architecture and domain-specific language for quantum computing. CoRR, abs/1402.4467, 2014
25. C. Developers, *Cirq*, 2021
26. T. Jones, A. Brown, I. Bush, S. Benjamin, Quest and high performance simulation of quantum computers. arXiv:1802.08032, 2018
27. M. Smelyanskiy, N.P.D. Sawaya, A. Aspuru-Guzik, qHiPSTER: The quantum high performance software testing environment. CoRR, abs/1601.07195, 2016
28. B. Villalonga, et al., A flexible high-performance simulator for verifying and benchmarking quantum circuits implemented on real hardware. npj Quantum Inf. **5**(1) 1–16 (2019)
29. *Forest SDK*, https://www.rigetti.com/systems, Accessed: 2020-07-22, 2020
30. T. Grurl, R. Kueng, J. Juß, R. Wille, Stochastic quantum circuit simulation using decision diagrams, in *Design, Automation and Test in Europe*, 2021

31. T. Grurl, J. Fuß, R. Wille, Considering decoherence errors in the simulation of quantum circuits using decision diagrams, in *Int'l Conf. on CAD*, pp. 140:1–140:7 (IEEE, 2020)
32. A. Barenco, et al., Elementary gates for quantum computation. Phys. Rev. A **52**(5), 3457–3467 (1995)
33. D. Maslov, On the advantages of using relative phase Toffolis with an application to multiple control Toffoli optimization. Phys. Rev. A **93**(2), 022311 (2016)
34. R. Wille, M. Soeken, C. Otterstedt, R. Drechsler, Improving the mapping of reversible circuits to quantum circuits using multiple target lines, in *Asia and South Pacific Design Automation Conf.*, 2013
35. P. Murali, J.M. Baker, A. Javadi-Abhari, F.T. Chong, M. Martonosi, Noise-adaptive compiler mappings for noisy intermediate-scale quantum computers, in *Int'l Conf. on Architectural Support for Programming Languages and Operating Systems*, 2019, pp. 1015–1029
36. M.Y. Siraichi, V.F. dos Santos, S. Collange, F.M.Q. Pereira, Qubit allocation, in *Proc. Int'l Symp. on Code Generation and Optimization*, 2018, pp. 113–125
37. A. Zulehner, A. Paler, R. Wille, An efficient methodology for mapping quantum circuits to the IBM QX architectures. IEEE Trans. CAD Integr. Circuits Syst. **38**(7), 1226–1236 (2019)
38. R. Wille, L. Burgholzer, A. Zulehner, Mapping quantum circuits to IBM QX architectures using the minimal number of SWAP and H operations, in *Design Automation Conf.*, 2019
39. G. Li, Y. Ding, Y. Xie, Tackling the qubit mapping problem for NISQ-era quantum devices, in *Int'l Conf. on Architectural Support for Programming Languages and Operating Systems*, 2019
40. A. Matsuo, W. Hattori, S. Yamashita, Reducing the overhead of mapping quantum circuits to IBM Q system, in *IEEE International Symposium on Circuits and Systems*, 2019
41. T. Itoko, R. Raymond, T. Imamichi, A. Matsuo, Optimization of quantum circuit mapping using gate transformation and commutation. Integration **70**, 43–50 (2020)
42. G. Vidal, C.M. Dawson, Universal quantum circuit for two-qubit transformations with three controlled-NOT gates. Phys. Rev. A **69**(1), 010301 (2004)
43. M.A. Nielsen, I.L. Chuang, *Quantum Computation and Quantum Information* (Cambridge University Press, 2010)
44. D. Janzing, P. Wocjan, T. Beth, Non-identity check is QMA-complete. Int. J. Quantum Inform. **3**(3), 463–473 (2005)
45. L. Burgholzer, R. Wille, Advanced equivalence checking for quantum circuits. IEEE Trans. CAD Integr. Circuits Syst. (2021)
46. L. Burgholzer, R. Raymond, R. Wille, Verifying results of the IBM Qiskit quantum circuit compilation flow, in *Int'l Conf. on Quantum Computing and Engineering*, 2020, pp. 356–365
47. J. Yuan, C. Pixley, A. Aziz, *Constraint-Based Verification* (Springer, 2006)
48. J. Bergeron, *Writing Testbenches using System Verilog* (Springer, 2006)
49. N. Kitchen, A. Kuehlmann, Stimulus generation for constrained random simulation, in *Int'l Conf. on CAD*, 2007, pp. 258–265
50. R. Wille, D. Große, F. Haedicke, R. Drechsler, SMT-based stimuli generation in the SystemC Verification library, in *Forum on Specification and Design Languages*, 2009
51. K. Laeufer, J. Koenig, D. Kim, J. Bachrach, K. Sen, RFUZZ: Coverage-directed fuzz testing of RTL on FPGAs, in *Int'l Conf. on CAD*, 2018
52. H.M. Le, D. Große, N. Bruns, R. Drechsler, Detection of hardware trojans in SystemC HLS designs via coverage-guided fuzzing, in *Design, Automation and Test in Europe*, 2019, pp. 602–605
53. L. Burgholzer, R. Kueng, R. Wille, Random stimuli generation for the verification of quantum circuits, in *Asia and South Pacific Design Automation Conf.*, 2021

Layout Synthesis for Near-Term Quantum Computing: Gap Analysis and Optimal Solution

Bochen Tan and Jason Cong

1 Introduction

The nature of quantum computing (QC) decides that it is much more error-prone than classical computing. The scalable and effective way to resolve this issue is encoding quantum information with quantum error correction (QEC) codes [16, 36], but current technology cannot produce a quantum processor with enough size and fidelity for QEC. In this chapter, we consider near-term QC without QEC. Note that the "near-term" formalism is by no means ephemeral. The early QEC-capable hardware will not be able to run any application on top of the QEC schemes because QEC has huge overheads. It is estimated that application-ready error-corrected quantum computers are at least 10 years away [21, 30], assuming many engineering challenges [1] are solved. Meanwhile, all the QC applications can only use the near-term formalism [32].

Hardware technology draws the upper bound of QC capability, while compilation determines if and how much we can utilize this capability. A QC compiler has to transform the input quantum program to satisfy constraints imposed by the quantum *architecture*. In addition, errors accumulate fast when scaling up QC without QEC, so the compiler should make best efforts to reduce errors. There has been over a decade of QC compilation research, as summarized in Sect. 2, so it is vital to quantitatively examine these existing tools and, if there is still a significant room, improve them.

In general, there are two types of constraints in QC compilation: *logic constraints* and *layout constraints*. Quantum programs are specified as a list of instructions, e.g., Fig. 1a, where each one is an operation on a set of *program qubits*. They can also be

B. Tan · J. Cong (✉)
University of California, Los Angeles, CA, USA
e-mail: bctan@cs.ucla.edu; cong@cs.ucla.edu

© The Author(s), under exclusive license to Springer Nature Switzerland AG 2023
R. O. Topaloglu (ed.), *Design Automation of Quantum Computers*,
https://doi.org/10.1007/978-3-031-15699-1_2

25

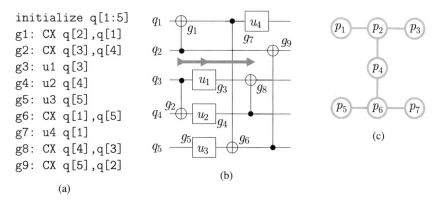

```
initialize q[1:5]
g1: CX q[2],q[1]
g2: CX q[3],q[4]
g3: u1 q[3]
g4: u2 q[4]
g5: u3 q[5]
g6: CX q[1],q[5]
g7: u4 q[1]
g8: CX q[4],q[3]
g9: CX q[5],q[2]
```

(a)

(b)

(c)

Fig. 1 Inputs of the layout synthesis problem: quantum program and coupling graph. (**a**) A quantum program consisting of 9 gates on 5 program qubits. (**b**) Circuit of the quantum program. Each horizontal wire is a program qubit. Time goes from left to right. The arrow means a dependency chain (g_2, g_3, g_8). (**c**) Coupling graph of "segment H on a Falcon processor" from IBM Quantum [20]. Each vertex is a physical qubit. Entangling two-qubit gates can only be applied to adjacent physical qubits

drawn as circuit diagrams like Fig. 1e where each wire represents a program qubit and each gate corresponds to an operation. A valid gate on n qubits is represented by a unitary matrix of dimension 2^n, e.g.,

$$H = \begin{bmatrix} \frac{\sqrt{2}}{2} & \frac{\sqrt{2}}{2} \\ \frac{\sqrt{2}}{2} & -\frac{\sqrt{2}}{2} \end{bmatrix}, U(\theta, \lambda, \phi) = \begin{bmatrix} \cos\left(\frac{\theta}{2}\right) & -e^{i\lambda}\sin\left(\frac{\theta}{2}\right) \\ e^{i\phi}\sin\left(\frac{\theta}{2}\right) & e^{i\lambda+i\phi}\cos\left(\frac{\theta}{2}\right) \end{bmatrix}, CX = \begin{bmatrix} 1 & 0 & 0 & 0 \\ 0 & 1 & 0 & 0 \\ 0 & 0 & 0 & 1 \\ 0 & 0 & 1 & 0 \end{bmatrix},$$

(1)

where H and U are *single-qubit gates*, and CX is a *two-qubit gate*. Moreover, U is a *programmable gate* with three parameters. We can tune the parameters to instantiate U to specific gates, e.g., U with $\theta = \pi/2$, $\lambda = \pi$, and $\phi = 0$ is just H. In our example, u_1 to u_4 are instances of U, each black dot is the first qubit for a CX, and each $\oplus$ is the second qubit for a CX. The logic constraint of QC hardware is specified as a *native gate set*, e.g., on IBM quantum computers, the set is $\{U, CX\}$ [20]. Gates not contained in this set, e.g., two-qubit gates other than CX, or gates on three or more qubits, have to be decomposed into a series of gates in the set to be executed on hardware. Fortunately, most near-term quantum algorithms are just written in single-qubit and two-qubit gates [11, 15, 25], and there are canonical decompositions of an arbitrary two-qubit gate into Us and CXs [31, 43] as displayed in Fig. 2. Other *important* multi-qubit gates in QC also have efficient decompositions [5].

An *entangling two-qubit gate* like CX is essential for QC. However, they are not available for all pairs of qubits. The layout constraints of a quantum processor are specified by a *coupling graph* $G = (P, E)$ like Fig. 1c where each vertex is a

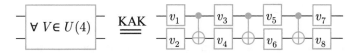

Fig. 2 KAK decomposition. For any V that is a 4-by-4 unitary matrix, the corresponding two-qubit gate can be decomposed into 3 CXs and 8 single-qubit gates

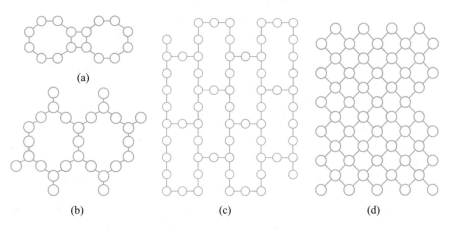

Fig. 3 Coupling graphs of some existing quantum processors. (**a**) Rigetti Aspen-4 [12]. (**b**) IBM Falcon [23]. (**c**) IBM Hummingbird [20]. (**d**) Google Sycamore [3]

physical qubit, and each edge means two-qubit gates can be applied to those two vertices. Thus, we need to map the program qubits in Fig. 1b to physical qubits in Fig. 1c so that the CX gates are on adjacent physical qubits. This is not always possible. In fact, the gates g_1, g_6, and g_9 fully connect the program qubits q_1, q_2, and q_5. No matter how we map the qubits, satisfying these three gates requires a triangle on the coupling graph, which does not exist in Fig. 1c. A way to resolve this issue is by inserting SWAP gates that changes the mapping in the middle of the circuit. However, SWAP gates bring error and may also increase the runtime, so a compiler needs to carefully decide when and where to insert them. To summarize, in the *layout synthesis* i.e., *qubit mapping* phase of compilation, program qubits are mapped to physical qubits, some circuit transformations, e.g., via SWAP gates, are performed, and all the gates are scheduled.

We believe that qubit mapping poses a more serious challenge than gate decomposition. This is because the native gates are decided by fundamental physics of the QC platform, but the coupling graphs have more degrees of design freedom. In fact, the native gate set of IBM quantum computers has not changed greatly since the beginning of their cloud QC service because the fundamental qubit technology is the same, but QC devices with very different coupling graphs have been introduced in the past a few years [20], e.g., Fig. 3b and c. With better understanding of the qubits, new coupling graphs are also being proposed and selected [18].

2   Previous Works

NP-completeness of several versions of the layout synthesis problem has been proven [7, 27, 37, 41]. Therefore, we can categorize previous works into heuristic ones and exact/optimal ones. The runtime of the latter scales exponentially in problem size because of the computational complexity of the problem.

In general, the heuristic works formulate layout synthesis as a search problem [2, 10, 26, 27, 37–39, 46–48]. In the search algorithms, the state is $(\mathcal{G}, \Pi)$ where $\mathcal{G}$ contains the gates that have been considered and Π is the current qubit mapping; the action leading to another state is either changing Π by SWAP(s) or appending some gates into $\mathcal{G}$ if they only act on adjacent qubit(s) under the current mapping; the cost of a Π-change is often evaluated by looking ahead a few more steps in the search tree. At the beginning of the search, $\mathcal{G}$ is just empty and there are a few different ways to find the initial mapping Π_0. References [27, 37, 38] use the earlier two-qubit gates to construct a program graph between program qubits and apply existing graph isomorphism algorithms from this program graph to the coupling graph. In [37], the program graph is additionally weighted by the number of two-qubit gates between this qubit pair. References [2, 10, 39, 46–48] start the search with some Π_0 and expand the search tree a few times. Then, they select the best mapping so far and use it as the real initial mapping. References [26] searches for the final mapping of the reversed program and uses it as Π_0 for the original program.

In theory, one can derive the optimal solution by fully expanding the search tree in the heuristic search approaches, but, in practice, the exact/optimal approaches formulate the layout synthesis into mathematical programming and apply a solver: [28, 40, 42, 45] use satisfiability modulo theories (SMT) solvers, [6, 29, 35] use integer programming (IP) solvers, and [44] uses temporal planners. To reduce runtime, many works compromise by slicing the program and only considering the next slice when inserting SWAPs [6, 28, 29, 35].

3   The Measure-Improve Methodology and Its Application in Classical Circuit Placement

Given the extensive amount of work on layout synthesis, a natural question is if these solutions are close to optimal. However, it remains challenging how to measure their optimality.

The layout synthesis problem summarized above is representative of many problems in design automation or, broadly, in computer science: the complexity is NP-hard, and they can be formulated into some kind of mathematical programming and solved with exponential runtime. To solve large instances of these problems, one approach is to accelerate the solver, often in a domain-specific way. Another approach is to develop heuristic methods that run faster but are not optimal. How do we evaluate these heuristics? A common way is using a set of representative applications as the benchmark and comparing the results by different heuristics. However, because of the complexity of the problem, we do not know the optimal result of these benchmarks, so we do not know how much room of improvement there is. If, after substantial research, the improvements are diminishing, the community may be in a dilemma: is it possible that the current heuristics are very close to optimal and further research will produce diminishing returns; or is there still significant room requiring fresh ideas and more efforts? We cannot be certain about both possibilities since deriving optimal solution for large instances takes astronomical time. In this case, it would be helpful if there are benchmarks with known optimal solution and the size of these benchmarks should be large enough to imitate real applications. With such benchmarks, we can measure the sub-optimality of the heuristics and improve them if there is still significant room.

One such benchmark set in classical circuit design is PEKO [9], *placement examples with known optimal*. Before placement, the circuit is represented as connected *modules* shown as C_i, $i \in [4]$, in Fig. 4a. The input of the problem is thus a *netlist* where each *net* connects two or more pins on different modules. In our example, there is a 2-pin net connecting C_1 and C_4, and a 4-pin net connecting all the modules. After placing the modules on the chip area, manufacturers need to implement the nets with wires, as shown in Fig. 4b. This example is not ideal since the total wirelength can be reduced if we place the modules closer together, e.g., by putting the modules at the four cells in the bottom right corner. Despite that placement is known to be NP-complete [33], the authors of [9] present a way of constructing placement examples with known optimal wirelength from locally optimal nets, as demonstrated in Fig. 4c. That is, one follows the net size distribution specification. For each net of size r, one connects pins from $\lceil \sqrt{r} \rceil \times \lceil \sqrt{r} \rceil$ adjacent modules. Since all the nets are among adjacent modules, the total wirelength cannot

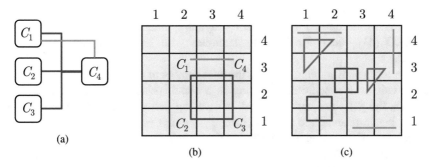

(a)   (b)   (c)

Fig. 4 The placement problem in classical circuit design. (**a**) Netlist, input of the placement problem. Each net connects some pins on modules $C_i, i \in [4]$. There is a 2-pin net (orange) between C_1 and C_4, and a 4-pin net (purple) connecting all modules. (**b**) A placement solution. The modules are placed to cells in the chip area. This solution is not ideal since the modules are placed far away, so longer wires are needed by the nets. (**c**) Construction of PEKO, placement example with known optimal [9]. The nets are all shortest possible, so the shortest wire length of the whole netlist is just the sum of each one

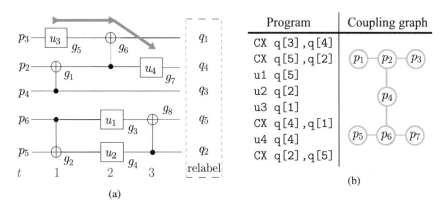

(a)

(b)

Fig. 5 Quantum mapping example with known optimal (QUEKO) [41]. (**a**) Constructing a program with optimal depth 3 from a physically realizable circuit. The orange arrow is a backbone of length 3: (g_5, g_6, g_7). The program is derived by relabeling the wires to program qubits. (**b**) A QUEKO benchmark consisting of a program $\mathcal{P}$ and a coupling graph G. The optimal depth T_O of mapping $\mathcal{P}$ to G is known by construction to be 3. The minimal SWAP count is 0

be reduced. Thus, we know the optimal wirelengths of PEKO benchmarks by construction. The PEKO benchmarks were used to measure optimality of leading placers at that time and showed 2X optimality gap, which spurred the community to invest more efforts into the placement problem. This led to a wirelength reduction equivalent to two generations of hardware scaling in Moore's law [34].

4 Measuring Optimality with QUEKO

To measure the optimality of existing layout synthesis solutions, we developed QUEKO [41]—quantum mapping examples with known optimal, inspired by PEKO.

The depth of a quantum circuit is the total number of time steps it takes to execute the circuit on a specific architecture. In a circuit prior to layout synthesis, e.g., Fig. 1b, the gates are not scheduled, but we can derive some lower bounds of its depth. For instance, there are three gates g_2, g_3, and g_8 subsequently acting on q_3. Since none of them can be executed at the same time, we need at least three time steps to run these gates. Moreover, we cannot change the order of execution, since g_3 depends on g_2, and g_8 depends on g_3. The maximal length of *dependency chains* like (g_2, g_3, g_8) is a lower bound for depth. Of course, it is not always possible to reach this lower bound because we may need to insert SWAPs in layout synthesis which could lengthen the dependency chains. This is indeed the case for the example of Fig. 1, since SWAP insertion is absolutely necessary, as we have explained in Sect. 1.

Table 1 Optimality gaps[a] measured by QUEKO benchmarks of feasible depths[b]

Compiler[c]	Small architecture[d] and sparse program[e]				Large architecture and dense program			
	10	20	30	40	10	20	30	40
JKU [48]	13X	11X	9.3X	8.7X	Process runs out of RAM (128GB)			
Cirq [10]	8.3X	9.3X	7.9X	9.2X	44X	45X	48X	47X
Qiskit [2]	5.4X	4.7X	4.8X	4.6X	12X	12X	11X	11X
t|ket⟩ [38]	1.04X	1.32X	1.30X	1.72X	1.82X	3.1X	3.7X	5.7X

[a] Optimality gap is defined as T/T_O where T_O is the optimal depth and T is the depth of the result produced by a compiler

[b] For each $T_O = 10, 20, 30$, and 40, we generated 10 QUEKO benchmarks and input them to the compilers being examined. The data shown above are geometric means of the 10 corresponding optimality gaps

[c] We used the Greedy router in Cirq 0.6.0, DenseLayout followed by StochasticSwap in Qiskit 0.14.1, Graph Placement followed by Route in t|ket⟩ 0.4.1

[d] The "small architecture" is Rigetti Aspen-4 (Fig. 3a) and the big one is Google Sycamore (Fig. 3d)

[e] The "sparse program" has the gate density of the Toffoli gate, and the dense one has that of the quantum supremacy experiment [3]

For physically realizable quantum circuits where all the two-qubit gates act on adjacent qubits, the maximal length of dependency chains is indeed the depth, not merely a lower bound. For instance, in Fig. 5a, one of the longest dependency chains is (g_5, g_6, g_7), which means the depth cannot be even lower than 3. We can construct a qubit mapping example with known optimal depth T_O by generating a physically realizable circuit with a *backbone* like (g_5, g_6, g_7) which is a dependency chain of length T_O. Building the backbone does not take too many gates, so we have another degree of freedom in QUEKO named *gate density*: how much of the spacetime is taken by idleness, single-qubit gates, and two-qubit gates. In Fig. 5a, there are three time steps and five qubits, so the spacetime volume of the circuit is 15. Each single-qubit gate takes one unit of volume, and each two-qubit gate takes two units. In Fig. 5a, the single-qubit gates $(g_5, g_3, g_4,$ and $g_7)$ take 4 out of 15 units of volume; the two-qubit gates $(g_1, g_2, g_6,$ and $g_8)$ take 8/15 volume; the reset 3/15 volume is idleness. Thus, the gate density for this circuit is (3/15, 4/15, 8/15). We relabel the physical qubits to program qubits to derive the quantum program that can be input to the compilers, as shown in the dashed box in Fig. 5a, e.g., p_3 is relabeled as q_1, so the single-qubit gate u_3 is on q_1 in Fig. 5b. If a compiler finds out the inverse of our relabeling, it will derive the original physically realizable circuit exactly.

In summary, given a coupling graph G, a depth T_O, and a gate density vector, we can construct a quantum program $\mathcal{P}$ with the known optimal depth T_O if mapped to G, as the demonstrated in Fig. 5b. We can pass the layout synthesis problem $(\mathcal{P}, G)$ to a compiler and check how far is the depth of its result, T, compared to the optimal depth T_O. The backbones in QUEKO benchmarks guarantee that T cannot be smaller than T_O. We can easily derive a depth-optimal solution by reading off the relabeling during QUEKO construction, e.g., for the problem in Fig. 5b, an optimal solution is $q_1 \mapsto p_3, q_4 \mapsto p_2, q_3 \mapsto p_4, q_5 \mapsto p_6,$ and $q_2 \mapsto p_5$ in Fig. 5a. Since this solution does not contain any SWAPs, QUEKO benchmarks also have known optimal number of SWAPs, which is 0.

With the help of QUEKO, we find that, despite over a decade long research in layout synthesis, the optimality gaps are still large, as shown in Table 1. The depth of results by compilers other than t|ket⟩ is at least 4.6X the optimal depth. The rightmost column corresponds to QUEKO benchmarks that have the coupling graph, optimal depth, and gate density similar to the quantum supremacy experiment [3], which means that quantum circuits of this size are feasible. On these benchmarks, even the t|ket⟩ result is 5.7X the optimal. These gaps indicate that there is still substantial room for improvements in layout synthesis.

5   SMT Formulation of the Optimal Layout Synthesis Problem

In this section, we would like to present a precise formulation of the layout synthesis problem based on a mathematical programming formulation, which can be solved later by using a satisfiable modulo theories (SMT) solver. We would like to reiterate a few assumptions. (1) The qubit mapping is from program qubits to physical qubits. It is not from basic entities in other problems, like fermions in chemistry simulation [25], to program qubits. (2) Conventionally, the logic constraints have been resolved before layout synthesis, so all the input gates can be executed on hardware; gate cancellation has also been done, so every input gate must be executed. (3) The architecture/coupling graph is fixed. This is true for existing general-purpose quantum processors [17, 19, 20, 22, 24]. However, the formulation can be extended to processors with programmable architectures [8], e.g., using neutral atom arrays [14].

5.1   Variables

A solution of the layout synthesis problem has three sets of variables: *mapping variables*, *schedule variables*, and *SWAP variables*. We shall demonstrate these variables in Fig. 6 which is a solution to the problem instance displayed in Fig. 1. At the end of the day, what we need to inform the hardware is when and where to execute each gate. This is shown as the circuit diagram in Fig. 6. Different from the diagram in Fig. 1b, in this diagram, each wire is a physical qubit, and the gates on the same column are executed at the same time step. On the left of Fig. 6, the physical qubits involved are colored black in the coupling graph.

The value of a mapping variable $\pi_{q,t}$ is the physical qubit where q is mapped at time t. For example, in the beginning q_5 is mapped to p_3, so $\pi_{q5,1} = p_3$. In Fig. 6, we annotated the program qubit on the wire where it is mapped.

The schedule variable t_g of each gate in the program is just the time step when it is executed, annotated below in Fig. 6, e.g., $t_{g5} = 1$ and $t_{g6} = 2$.

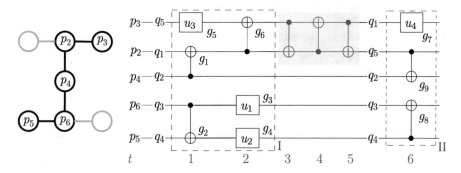

Fig. 6 A valid solution of the layout synthesis problem. On the left, the physical qubits involved are colored black in the coupling graph. On the right, each wire is a physical qubit. The time is shown below. Vertically aligned gates are executed at the same time step. The three green-shaded CX gates consist of a SWAP gate. The initial qubit mapping is shown before step 1, and the mapping after the SWAP is shown before step 6. The circuit can also be seen as two coarse-grain time steps (dashed boxes I and II) separated by a transition consisting of a SWAP on p_2 and p_3.

The SWAP variables $\sigma_{e,t}$ are binary variables that evaluate to 1 if and only if there is a SWAP gate finishing at time t on edge e. There is only one SWAP in Fig. 6, so the only non-zero SWAP variable is $\sigma_{(p_2,p_3),5}$.

We can set any function of the above variables as the objective. Three common objectives are straightforward: *depth*, the number of SWAPs, and *fidelity*. The depth of a quantum circuit is the maximum of all the schedule variables. The number of SWAP gates inserted is the sum of all the SWAP variables. The fidelity of the circuit, assuming a stochastic error model, is the product of all the individual gate fidelity, which can be input as the weights of vertices and edges in the coupling graph. There are four multiplicative terms in the fidelity expression: single-qubit gate fidelity, two-qubit gate fidelity, measurement fidelity, and SWAP fidelity.

5.2 Constraints

There has to be many constraints on the variable assignments in order for a valid solution. We can categorize them into four groups: connections, dependencies, no overlaps, and mapping transformations.

Connections Supposed that a two-qubit gate g at time t_g acts on program qubits q and q'. Then, there should be an edge between π_{q,t_g} and π_{q',t_g}. Otherwise, the physical qubits are not adjacent on the coupling graph and the two-qubit gate cannot be executed. For instance, for gate g_1 acting on q_1 and q_2,

$$t_{g_1} == 1 \implies (\pi_{q_1,1}, \pi_{q_2,1}) \in E, \tag{2}$$

where E is the edge set of the coupling graph.

Dependencies We have introduced the notion of dependency in Sec 4. In general, if two gates subsequently act on the same qubit, the order between them in the input program must be respected, e.g., since g_6 acts on q_5 after g_5,

$$t_{g_6} > t_{g_5}. \tag{3}$$

Note that with domain knowledge, we may relax some of these constraints, as specified later in Sect. 6.2.

No Overlaps Each qubit at each time step can only be involved in one gate, so we need to make sure that there are no overlaps between any gates including the SWAPs we inserted. For instance, $\sigma_{(p_2,p_3),6}$ cannot be 1 since, if so, the SWAP would finish on (p_2, p_3) at time 6 and overlap with g_7. This case is ruled out by constraints like

$$t_{g_7} == 6 \ \wedge \ \left(\pi_{q_1,6} == p_2 \ \vee \ \pi_{q_1,6} == p_3\right) \ \Rightarrow \ \sigma_{(p_2,p_3),6} == 0. \tag{4}$$

Mapping Transformations After a SWAP gate finishes, the qubit mapping should be transformed, e.g., the mapping of q_1 and q_5 exchanged at time 6, i.e.,

$$\pi_{q_5,5} == p_3 \wedge \pi_{q_1,5} == p_2 \wedge \sigma_{(p_2,p_3),5} == 1 \ \Rightarrow \ \pi_{q_1,6} == p_3, \ \pi_{q_5,6} == p_2. \tag{5}$$

On the other hand, if a qubit is not involved in any SWAP gates finishing at time t, its mapping variable should remain the same as time $t - 1$, e.g., for q_3 and $t = 3$,

$$\pi_{q_3,2} == p_6 \ \wedge \ \sigma_{(p_4,p_6),2} == 0 \ \wedge \ \sigma_{(p_6,p_5),2} == 0 \ \Rightarrow \ \pi_{q_3,3} == p_6. \tag{6}$$

We took a descriptive approach in this section, interested readers can refer to [40] for details. In total, there are $O(NTL)$ constraints where N is the number of physical qubits, T is the total depth, and L is the total number of gates.

6 Closing the Gap with OLSQ—Optimal Layout Synthesis for Quantum Computing

Upon the revelation of optimality gaps by QUEKO, we set out to close these gaps. As a result, we have formulated layout synthesis problem in Sect. 5. The variables and constraints are compatible with a kind of mathematical programming model named SMT, satisfiability modulo theories. Thus, we use an existing SMT solver, z3 [13], to derive the layout synthesis solutions optimally.

In terms of formulation, the main contribution of OLSQ is reducing the number of variables. In previous works like [45], there is a binary variable $x_{\Pi,t}$ at each time step t for a possible qubit mapping $\Pi : Q \rightarrow P$. For instance, the initial mapping in Fig. 6 is $\Pi_0: q_1 \mapsto p_2, q_2 \mapsto p_4, q_3 \mapsto p_6, q_4 \mapsto p_5$, and $q_5 \mapsto p_3$. Thus, $x_{\Pi_0,0} = 1$ and $x_{\Pi,0} = 0$ for any $\Pi \neq \Pi_0$. Note that there are exponentially many possible mappings with respect to the number of qubits, so there are exponentially many variables. In contrast, as Sect. 5.1 has demonstrated, OLSQ captures the mapping with $\pi_{q,t}$ variables and updates the mapping with the SWAP variables $\sigma_{e,t}$. For each time step, there are as many mapping variables as program qubits, *not* as many as possible qubit mappings. The number of SWAP variables is just the number of edges in the coupling graph, which is also linear in the number of physical qubits. Overall, OLSQ only has linearly many variables in the number of qubits, which is a huge improvement over previous works.

We implemented OLSQ in Python 3 and open-sourced the package[1] with BSD-3-Clause license. A user initializes OLSQ by choosing whether to use the transition mode, and the objective: depth, the number of SWAPs, or fidelity. Afterwards, the user inputs information about the hardware with setdevice(): the number of physical qubits, the edges, how many time steps a SWAP takes. If using fidelity as the objective, the fidelity of individual gates should be input at this stage. Then, the user should input the quantum program with setprogram(). Under the assumptions we introduced in the beginning of Sect. 5, OLSQ only needs a list of tuples to represent the program, each tuple for a gate. If it is a single-qubit gate, then the tuple only has one element, which is the index of the involved program qubit; if it is a two-qubit gate, then there are two elements in the tuple. Finally, the user can execute the solve() method to acquire the solution. The depth that OLSQ is currently trying will be printed on the screen while it calls z3 solver [13] to solve the SMT model corresponding to this depth. If there is a solution, the quantum circuit after layout synthesis would be returned; otherwise, OLSQ increases the depth and try again.

6.1 Speeding Up OLSQ with the Transition Mode

Because of the more efficient formulation, OLSQ shows better scaling in runtime compared to previous work with exact formulation [40]. However, because of the complexity of the problem, the runtimes are still long. Thus, we consider techniques that can accelerate the solving process with some sacrifice on optimality, which leads to the *transition mode*.

In Fig. 6, there are 6 time steps, but the mapping only changes once. So, many mapping variables, e.g., the ones for time steps 2 to 5, take the same value in the previous time step. We would have much less variables if we only keep the mapping

---

[1] https://github.com/UCLA-VAST/OLSQ.

variables when the mapping changes. In another perspective, this can be seen as if each gate has a "coarse-grain" schedule variable. Between the gates scheduled to coarse-grain time t and $t + 1$, there is a transition, which is a set of non-overlapping SWAPs. In Fig. 6, the two dashed boxes are two coarse-grain time steps I and II. The transition between them consists of the SWAP on (p_2, p_3).

To implement the transition mode, we just need to revise the OLSQ formulation by: 1) relaxing the $>$ in dependency constraints like Eq. 3 to $\geq$, 2) changing the duration of SWAP gates to 1, and 3) removing overlap constraints like Eq. 4 since in the transition-based model, the SWAP gates are part of transitions and will not interfere with the other gates. Note that the solver decides which gates go into which coarse-grain time step, so the transition-based (TB-)OLSQ is different from cutting the circuit beforehand and solving the sub-circuits separately. Thus, the results by TB-OLSQ are still much better than heuristics, as shown in Table 2, while achieving over 400X speedup compared to the original OLSQ on the benchmarks used in [40]. This is because the number of transition, $\tilde{T}$, is often much smaller than the circuit depth, T, so the number of variables $O(N\tilde{T} + L)$ and the number of constraints $O(\tilde{T}NL)$ also become much smaller than those of the original OLSQ.

6.2   Exploring Larger Solution Space

Apart from acceleration, we can also improve the solution quality given domain knowledge. One example of this is dropping some dependency constraints. In general, we cannot neglect these constraints for the correctness of results. However, for an important family of circuits named quantum approximate approximation algorithm, QAOA [15], there are many dependency constraints that can be dropped without consequences, because the ZZ gates inside QAOA are commutable, as illustrated in Fig. 7. This is the key factor of the improvements we see in Table 2: even there is a little sacrifice of optimality by using the transition mode, the depths and the numbers of SWAPs are at least halved compared to leading heuristics.

Table 2 Layout synthesis results of QAOA programs for 3-regular graphs[a]

QAOA size	8		10		12		14	
Compiler[b]	Depth	#SWAP	Depth	#SWAP	Depth	#SWAP	Depth	#SWAP
t|ket) [38]	13	6	15	7	20	13	22	14
SABRE [26]	11	5	15	9	13	9	20	11
TB-OLSQ [40]	5	3	6	5	6	5	7	4
OLSQ-GA [42]	4	0	3	0	4	0	4	0

[a] The whole setting is from [4], a leading experimental QAOA work: the QAOA benchmarks are generated from random 3-regular graphs of size 8, 10, 12, and 14, and the coupling graph is part of Google Sycamore. We only use one iteration of QAOA

[b] We used `pytket-cirq` 0.16.0, the Cirq integration of t\|ket); SABRE as integrated in Qiskit 0.27.0

$$ZZ(\gamma) = \mathrm{diag}(e^{-i\gamma},\ e^{i\gamma},\ e^{i\gamma},\ e^{-i\gamma})$$
$$ZZ(\gamma)\otimes I = \mathrm{diag}(e^{-i\gamma},\ e^{-i\gamma},\ e^{i\gamma},\ e^{i\gamma},$$
$$e^{i\gamma},\ e^{i\gamma},\ e^{-i\gamma},\ e^{-i\gamma})$$
$$I\otimes ZZ(\gamma) = \mathrm{diag}(e^{-i\gamma},\ e^{i\gamma},\ e^{i\gamma},\ e^{-i\gamma},$$
$$e^{-i\gamma},\ e^{i\gamma},\ e^{i\gamma},\ e^{-i\gamma})$$

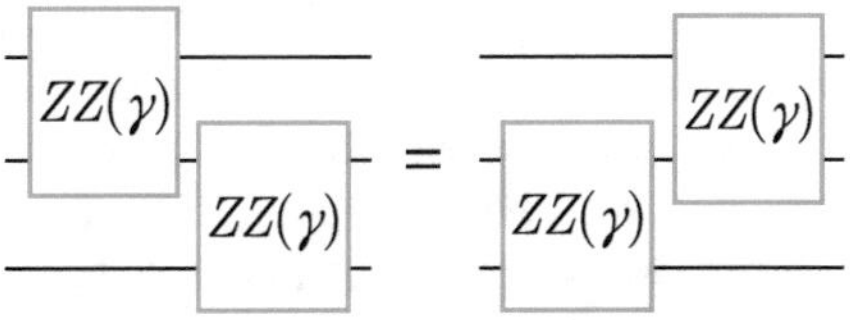

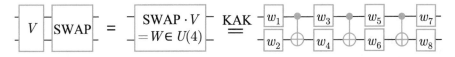

Fig. 7 Commutation of ZZ gates. $ZZ(\gamma)\otimes I$ means applying $ZZ(\gamma)$ gate on the upper two qubits while doing nothing on the bottom qubit. Since $ZZ(\gamma)\otimes I$ and $I\otimes ZZ(\gamma)$ are both diagonal, they are commutable

Fig. 8 A SWAP gate absorbed by gate V. In the first step, we compute the matrix product W of SWAP and V; in the second step, we apply KAK decomposition to W

Another example of exploring larger solution space is the technique of gate absorption, resulting in OLSQ-GA [42] where we combine layout synthesis with the synthesis of programmable two-qubit gates using KAK decomposition previously shown in Fig. 2. If a SWAP gate is directly after another two-qubit gate, e.g., in a QAOA circuit or other important circuits for chemistry [25] or machine learning [11], we can combine these two gates by computing the matrix product of them and synthesis this product, as illustrated by Fig. 8. The cost of implementing the gate induced by the product is much less than the cost of implementing the original two gates separately. The "absorption" of SWAP into other gates can be formulated by a set of absorbed SWAP variables $\alpha_{e,t}$ that behaves similarly to $\sigma_{e,t}$ variables, except bundled to other gates by constraints like

$$\alpha_{(p,p'),t} == 1 \ \Rightarrow\ \exists g\ \text{s.t.}\ t_g == t\ \wedge\ \pi_{q,t} == p\ \wedge\ \pi_{q',t} == p', \qquad (7)$$

where two-qubit gate g acts on program qubit q and q'. The results of OLSQ-GA are displayed in the bottom row of Table 2. Compared to TB-OLSQ, the depths decrease further, and more surprisingly, all the SWAPs are absorbed for this set of QAOA benchmarks, so there are no explicit SWAPs in the OLSQ-GA solutions.

7 Conclusion and Future Directions

In this chapter, we apply the measure-improve methodology, which has been successful in classical circuit placement, to the layout synthesis problem which is the center of compilation for near-term QC. We reveal quite large optimality gaps using QUEKO and aim to close these gaps by more efficient formulation, OLSQ, and more customized formulation OLSA-GA.

As future directions, (1) on the "measure" end, we plan to construct optimality benchmarks that are more similar to real QC applications; (2) on the "improve" end, we plan to accelerate the solving further by a top-down decomposition of problem instances, or a bottom-up approach using pre-computed optimal circuit library; 3) with an almost optimal compiler that has feasible runtime, we can evaluate quantum architecture designs without possible bias of the heuristics in compilation.

Acknowledgments This work is partially supported by the Center for Domain-Specific Computing Industrial Partnership Program.

References

1. C.G. Almudever et al., The engineering challenges in quantum computing, in *Design, Automation & Test in Europe Conference & Exhibition (DATE), 2017*. (IEEE, Lausanne, Switzerland, 2017), pp. 836–845. https://doi.org/10.23919/DATE.2017.7927104
2. M.S. Anis et al., Qiskit: An open-source framework for quantum computing (2021). [Online]. Available: https://doi.org/10.5281/zenodo.2573505
3. F. Arute et al., Quantum supremacy using a programmable superconducting processor. Nature **574**(7779), 505–510 (2019). arXiv:quant-ph/1910.11333. https://doi.org/10.1038/s41586-019-1666-5
4. F. Arute et al., Quantum approximate optimization of non-planar graph problems on a planar superconducting processor. Nature Physics **17**(3), 332–336 (2021). arXiv:quant-ph/2004.04197
5. A. Barenco, C.H. Bennett, R. Cleve, D.P. DiVincenzo, N. Margolus, P. Shor, T. Sleator, J.A. Smolin, H. Weinfurter, Elementary gates for quantum computation. Phys. Rev. A **52**(5), 3457–3467 (1995). https://doi.org/10.1103/PhysRevA.52.3457
6. D. Bhattacharjee, A.A. Saki, M. Alam, A. Chattopadhyay, S. Ghosh, MUQUT: Multi-constraint quantum circuit mapping on NISQ computers: Invited paper, in *2019 IEEE/ACM International Conference on Computer-Aided Design (ICCAD)* (IEEE, Westminster, CO, USA, 2019), pp. 1–7, https://doi.org/10.1109/ICCAD45719.2019.8942132
7. A. Botea, A. Kishimoto, R. Marinescu, On the complexity of quantum circuit compilation, in *Proceedings of the 11th Annual Symposium on Combinatorial Search* (AAAI Press, 2018), p. 5
8. S. Brandhofer, H.P. Büchler, I. Polian, Optimal mapping for near-term quantum architectures based on Rydberg atoms, in *2021 IEEE/ACM International Conference on Computer-Aided Design (ICCAD)*, 2021. arXiv:quant-ph/2109.04179
9. C.-C. Chang, J. Cong, M. Romesis, M. Xie, Optimality and scalability study of existing placement algorithms. IEEE Trans. Comput. Aided Des. Integr. Circuits Syst. **23**(4), 537–549 (2004). https://doi.org/10.1109/TCAD.2004.825870
10. Cirq Developers, Cirq (2021, Aug.). See full list of authors on GitHub: https://github.com/quantumlib/Cirq/graphs/contributors. [Online]. Available: https://doi.org/10.5281/zenodo.5182845
11. I. Cong, S. Choi, M.D. Lukin, Quantum convolutional neural networks. Nature Physics **15**(12), 1273–1278 (2019). arXiv:quant-ph/1810.03787. https://doi.org/10.1038/s41567-019-0648-8
12. A. Cornelissen, J. Bausch, A. Gilyén, Scalable benchmarks for gate- based quantum computers (2021). arXiv:quant-ph/2104.10698
13. L. de Moura, N. Bjørner, Z3: An efficient SMT solver, in *Tools and Algorithms for the Construction and Analysis of Systems*, ser. Lecture Notes in Computer Science, ed. by C.R. Ramakrishnan, J. Rehof (Springer, Berlin, Heidelberg, 2008), pp. 337–340. https://doi.org/10.1007/978-3-540-78800-3_24

14. S. Ebadi et al., Quantum phases of matter on a 256-atom programmable quantum simulator. Nature **595**(7866), 227–232 (2021). https://doi.org/10.1038/s41586-021-03582-4
15. E. Farhi, J. Goldstone, S. Gutmann, A quantum approximate optimization algorithm (2014). arXiv:quant-ph/1411.4028
16. A.G. Fowler, M. Mariantoni, J.M. Martinis, A.N. Cleland, Surface codes: Towards practical large-scale quantum computation. Phys. Rev. A **86**(3), 032324 (2012). arXiv:quant-ph/1208.0928. https://doi.org/10.1103/PhysRevA.86.032324
17. Google Quantum AI, Quantum computer datasheet (2021). [Online]. Available: https://quantumai.google/hardware/datasheet/weber.pdf
18. J.B. Hertzberg, R.O. Topaloglu, Quantum circuit topology selection based on frequency collisions between qubits. US Patent US20 200 401 925A1 (2020). [Online]. Available: https://patents.google.com/patent/US20200401925A1/en/
19. Honeywell, Honeywell sets new record for quantum computing performance (2020). [Online]. Available: https://www.honeywell.com/us/en/news/2021/03/honeywell-sets-new-record-for-quantum-computing-performance
20. IBM Quantum Processor, [Online]. Available: https://quantum-computing.ibm.com/services/docs/services/manage/systems/processors
21. IBM, 5 things to know about the IBM roadmap to scaling quantum technology (2020). [Online]. Available: https://newsroom.ibm.com/5-Things-About-IBM-Roadmap-to-Scale-Quantum-Technology
22. IONQ, Ionq (2020). [Online]. Available: https://ionq.com/technology
23. P. Jurcevic et al., Demonstration of quantum volume 64 on a superconducting quantum computing system. Quantum Sci. Technol. **6**(2), 025020 (2021). https://doi.org/10.1088/2058-9565/abe519
24. P.J. Karalekas, N.A. Tezak, E.C. Peterson, C.A. Ryan, M.P. da Silva, R.S. Smith, A quantum-classical cloud platform optimized for variational hybrid algorithms. Quantum Sci. Technol. **5**(2), 024003 (2020). arXiv:quant-ph/2001.04449. https://doi.org/10.1088/2058-9565/ab7559
25. I.D. Kivlichan, J. McClean, N. Wiebe, C. Gidney, A. Aspuru-Guzik, G.K.-L. Chan, R. Babbush, Quantum simulation of electronic structure with linear depth and connectivity. Phys. Rev. Lett. **120**(11), 110501 (2018). arXiv:quant-ph/1711.04789. https://doi.org/10.1103/PhysRevLett.120.110501
26. G. Li, Y. Ding, Y. Xie, Tackling the qubit mapping problem for NISQ-era quantum devices, in *Proceedings of the Twenty-Fourth International Conference on Architectural Support for Programming Languages and Operating Systems - ASPLOS '19*. (ACM Press, Providence, RI, USA, 2019), pp. 1001–1014. arXiv:cs.ET/1809.02573. https://doi.org/10.1145/3297858.3304023
27. D. Maslov, S.M. Falconer, M. Mosca, Quantum circuit placement. IEEE Trans. Comput. Aided Des. Integr. Circuits Syst. **27**(4), 752–763 (2008). arXiv:2002.09783. https://doi.org/10.1109/TCAD.2008.917562
28. P. Murali, N.M. Linke, M. Martonosi, A.J. Abhari, N.H. Nguyen, C.H. Alderete, Full-stack, real-system quantum computer studies: Architectural comparisons and design insights, in *Proceedings of the 46th International Symposium on Computer Architecture - ISCA '19* (ACM Press, Phoenix, Arizona, 2019), pp. 527–540. arXiv:quant-ph/1905.11349. https://doi.org/10.1145/3307650.3322273
29. G. Nannicini, L.S. Bishop, O. Gunluk, P. Jurcevic, Optimal qubit assignment and routing via integer programming. arXiv:quant-ph/2106.06446
30. H. Neven, Keynote in Google Quantum Summer Symposium (2020). [Online]. Available: https://youtu.be/HgQOPhNCct0
31. E.C. Peterson, G.E. Crooks, R.S. Smith, Fixed-depth two-qubit circuits and the monodromy polytope. Quantum **4**, 247 (2020). arXiv:1904.10541. https://doi.org/10.22331/q-2020-03-26-247
32. J. Preskill, Quantum computing in the NISQ era and beyond. Quantum **2**, 79 (2018). arXiv:quant-ph/1801.00862. https://doi.org/10.22331/q-2018-08-06-79

33. S. Sahni, A. Bhatt, The complexity of design automation problems, in *Proceedings of the 17th Design Automation Conference*, ser. DAC '80 (Association for Computing Machinery, New York, NY, USA, 1980), pp. 402–411. https://doi.org/10.1145/800139.804562

34. Semiconductors Research Corporation, 'Huge opportunity' in IC design optimization gained by Semiconductor Research Corporation (2007). National Science Foundation: CAD innovation could save industry billions. [Online]. Available: https://www.src.org/newsroom/press-release/2007/41/

35. A. Shafaei, M. Saeedi, M. Pedram, Qubit placement to minimize communication overhead in 2D quantum architectures, in *2014 19th Asia and South Pacific Design Automation Conference (ASP-DAC)* (IEEE, Singapore, 2014), pp. 495–500. https://doi.org/10.1109/ASPDAC.2014.6742940

36. P.W. Shor, Scheme for reducing decoherence in quantum computer memory. Phys. Rev. A **52**(4), R2493–R2496 (1995). https://doi.org/10.1103/PhysRevA.52.R2493

37. M.Y. Siraichi, V.F. dos Santos, S. Collange, F.M.Q. Pereira, Qubit allocation, in *Proceedings of the 2018 International Symposium on Code Generation and Optimization - CGO 2018* (ACM Press, Vienna, Austria, 2018), pp. 113–125. https://doi.org/10.1145/3168822

38. S. Sivarajah, S. Dilkes, A. Cowtan, W. Simmons, A. Edgington, R. Duncan, t|ket〉: A retargetable compiler for NISQ devices. Quantum Sci. Technol. (2020). arXiv:quant-ph/2003.10611. https://doi.org/10.1088/2058-9565/ab8e92

39. R.S. Smith, E.C. Peterson, M.G. Skilbeck, E.J. Davis, An open-source, industrial-strength optimizing compiler for quantum programs. Quantum Sci. Technol. **5**(4), 044001 (2020). https://doi.org/10.1088/2058-9565/ab9acb

40. B. Tan, J. Cong, Optimal layout synthesis for quantum computing, in *2020 IEEE/ACM International Conference on Computer-Aided Design (ICCAD)*, ser. ICCAD '20 (Association for Computing Machinery, Virtual Event, USA, 2020). arXiv:quant-ph/2007.15671. https://doi.org/10.1145/3400302.3415620

41. B. Tan, J. Cong, Optimality study of existing quantum computing layout synthesis tools. IEEE Trans. Comput. (2020). arXiv:quant-ph/2002.09783. https://doi.org/10.1109/TC.2020.3009140

42. B. Tan, J. Cong, Optimal qubit mapping with simultaneous gate absorption, in *2021 IEEE/ACM International Conference on Computer-Aided Design (ICCAD)*, ser. ICCAD '21 (Association for Computing Machinery, Munich, Germany, 2021). arXiv:cs.ET/2109.06445

43. F. Vatan, C. Williams, Optimal quantum circuits for general two-qubit gates. Phys. Rev. A **69**(3), 032315 (2004). https://doi.org/10.1103/PhysRevA.69.032315

44. D. Venturelli, M. Do, E. Rieffel, J. Frank, Compiling quantum circuits to realistic hardware architectures using temporal planners. Quantum Sci. Technol. **3**(2), 025004 (2018). https://doi.org/10.1088/2058-9565/aaa331

45. R. Wille, L. Burgholzer, A. Zulehner, Mapping quantum circuits to IBM QX architectures using the minimal number of SWAP and H operations, in *Proceedings of the 56th Annual Design Automation Conference 2019 on - DAC '19* (ACM Press, Las Vegas, NV, USA, 2019), pp. 1–6. arXiv:quant-ph/1907.02026. https://doi.org/10.1145/3316781.3317859

46. C. Zhang, A.B. Hayes, L. Qiu, Y. Jin, Y. Chen, E.Z. Zhang, Time-optimal Qubit mapping, in *Proceedings of the 26th ACM International Conference on Architectural Support for Programming Languages and Operating Systems* (ACM, Virtual USA, 2021), pp. 360–374. https://doi.org/10.1145/3445814.3446706

47. A. Zulehner, R. Wille, Compiling SU(4) quantum circuits to IBM QX architectures, in *Proceedings of the 24th Asia and South Pacific Design Automation Conference on - ASPDAC '19* (ACM Press, Tokyo, Japan, 2019), pp. 185–190. https://doi.org/10.1145/3287624.3287704

48. A. Zulehner, A. Paler, R. Wille, Efficient mapping of quantum circuits to the IBM QX architectures, in *2018 Design, Automation & Test in Europe Conference & Exhibition (DATE)* (IEEE, Dresden, Germany, 2018), pp. 1135–1138. arXiv:quant-ph/1712.04722. https://doi.org/10.23919/DATE.2018.8342181

Towards Efficient Superconducting Quantum Processor Architecture Design

Gushu Li (iD)**, Yufei Ding, and Yuan Xie**

1 Introduction

As a promising computational paradigm, Quantum Computing (QC) has been rapidly growing in the last two decades and found its strong potential in many important areas, including machine learning [1, 2], chemistry simulation [3, 4], etc. In particular, the superconducting quantum circuit [5] has become one of the most promising technique candidates for building QC systems [6–8] due to the ever-increasing qubit coherence time, individual qubit addressability, fabrication technology scalability, etc. Towards efficient superconducting quantum circuit based QC systems, significant research has recently been conducted, ranging from compiler optimization [9, 10] to periphery control hardware support [11, 12] and device innovation [13, 14].

Despite these system optimizations, the performance of a superconducting quantum processor is still highly limited by the amount of computational resources on it. Researchers have been trying to integrate more qubits and qubit connections on one superconducting quantum processor substrate. For example, IBM's first superconducting quantum chip on the cloud has 5 qubits with 6 qubit connections, while its latest published chip has 20 qubits with 37 qubit connections [15]. Increasing the number of physical qubits on a superconducting quantum processor allows programs with more logical qubits to be executed. Denser qubit connections can increase the overall chip performance by reducing the overhead of qubit mapping and routing [16–19].

Nevertheless, more qubits and qubit connections will, unfortunately, increase the probability of defect occurrence on a chip, leading to lower yield rate and blocking future development of larger-scale superconducting quantum processors.

G. Li (✉) · Y. Ding · Y. Xie
University of California, Santa Barbara, CA, USA
e-mail: gushuli@ucsb.edu; yufeiding@ucsb.edu; yuanxie@ucsb.edu

© The Author(s), under exclusive license to Springer Nature Switzerland AG 2023
R. O. Topaloglu (ed.), *Design Automation of Quantum Computers*,
https://doi.org/10.1007/978-3-031-15699-1_3

For example, the yield rate of a 17-qubit chip can be lower than 10% under state-of-the-art technology [20]. Such a low yield rate comes from *frequency collision*, a unique defect on superconducting quantum processors [20, 21]. The frequencies of physically connected qubits may "collide" with each other when their values satisfy some specific conditions. More qubit connections naturally increase the probability of frequency collision and lower the yield rate.

To optimize both the yield rate and performance would be desirable, but it is difficult in general due to the inherent trade-off between these two objectives. Most previous efforts on them are direct device-level improvement [13, 14, 22, 23], while little attention has been given to the architectural design of a superconducting quantum processor. This study fills the gap by exploring the possibility of efficient *application-specific architecture design* to reach an optimized balance between yield rate and performance. Our vision is that an array of QC accelerators, each of which is tailored to a specific application, is much more likely to be adopted in the near term where computational resources are still limited before we can reach a universal quantum computer (i.e., one quantum computer that runs all kinds of quantum programs). Our design shares the same high-level spirit with the hardware architecture designs in classical computing (e.g., machine learning [24, 25], graph processing [26, 27]), but faces different scenarios because both the program patterns and the hardware design space are different in QC.

In particular, we highlight two key challenges to be addressed before the application-specific principle can be applied in superconducting quantum processor design. **First**, we need to identify and abstract the computational pattern of quantum programs that can guide the hardware architecture design. Prior quantum program analysis studies [28–33] mainly focused on software or compiler optimization and cannot extract appropriate information for hardware architecture optimization. **Second**, the abstracted computational pattern must give guidance to efficient architectural designs, which employ fewer computational resources with physical constraints satisfied to achieve both high yield rate and performance. Existing superconducting quantum processor design schemes cannot handle such irregular/complicated application-specific architecture design tasks [34–38].

To overcome these two challenges, we design a systematic design flow to automatically generate efficient superconducting quantum processor architecture designs for different quantum programs (shown in Fig. 1). We first identify two key computational patterns in quantum programs, *coupling degree list* and *coupling strength matrix*. A profiler is built to automatically extract them from an input quantum program. Both of them are critical to the program performance and hardware yield rate, and thus optimizing their underlying architecture support can potentially achieve a better balance between the performance and yield rate. We then propose an architecture design flow, which comes with three key subroutines, *layout design*, *bus selection*, and *frequency allocation*. Each subroutine focuses on different hardware resources and must cooperate with corresponding profiling results and physical constraints. We further propose an array of heuristics to ensure the scalability and effectiveness of the architecture search process. Empirical studies

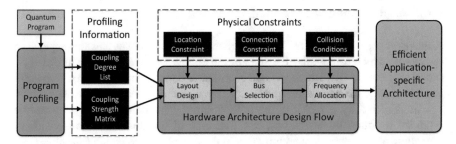

Fig. 1 Overview of the proposed architecture design flow

show that these heuristics can find "near-optimal" solution in the reduced search space.

In summary, this chapter makes the following contributions:

- We identify the optimization opportunity from the architecture level to push forward the balance between performance and hardware yield rate for superconducting QC processors.
- We formalize an end-to-end design flow, equipped with a set of novel algorithmic primitives, to automatically generate a series of application-specific architectural designs under different hardware resource limits.
- Comprehensive experiments show that our design flow could outperform known general-purpose designs with better Pareto-optimal results, e.g., magnitudes of yield improvement with negligible performance loss.

2   Background

In this section, we will introduce the necessary QC basics for understanding the following program profiling and superconducting quantum processor architecture design.

2.1   QC Program Basics

A quantum program can be represented in the well adopted quantum circuit model [39, 40]. We will start from the basic components in a quantum circuit and then illustrate how they compose a quantum circuit.

2.1.1   Logical Qubit and Quantum Operation

A quantum program consists of some logical qubits as variables and some quantum operations which can modify the state of the qubits. Qubit is the basic information processing unit in QC, which has two basis states denoted as $|0\rangle$ and $|1\rangle$. One qubit can be not only the basis states themselves but also their linear combinations which can be depicted by a vector in the Hilbert space. The state of the qubits can be modified by quantum operations. The first type of quantum operation is unitary operation, also known as quantum gates in the circuit model, which can implement a unitary transformation on the qubit state. Quantum gates can be applied on single qubit or multiple qubits. The second type is measurement operation, which forces the qubits to collapse to basis states.

2.1.2   Quantum Circuit

Quantum circuit is a model of QC in which the computation is a sequence of quantum gates and measurement operations. The state of the qubits is first initialized and then manipulated by a sequence of operations. Single-qubit gates and measurement operations are applied on individual qubits while two-qubit gates are applied on two logical qubits. It has been proved that any multi-qubit gate can be decomposed into a series of single-qubit gates and CNOT gates (a specific two-qubit gate) [41]. This is also the basic gate set directly supported on IBM's devices. As a result, it is assumed that the quantum circuit has been decomposed and gates with three or more qubits are not considered.

2.2   Superconducting Quantum Circuit Basics

All the qubits and quantum operations in a quantum circuit must be implemented in a real physical QC system to execute the program. Here we focus on superconducting quantum processors with fixed-frequency Josephson-junction-based transmon qubits [13] and all-microwave cross-resonance two-qubit gates [42] adopted by IBM [36].

2.2.1   Physical Qubit and Frequency

Figure 2 shows the physical circuit and energy levels of a transmon qubit [13]. Due to the nonlinearity of the Josephson junction, the gaps between the energy levels in this quantum anharmonic oscillator are different, which allows us to use the ground state $|0\rangle$ and the first-excited state $|1\rangle$ as the computational basis without populating other states. Suppose the energy gap between $|0\rangle$ and $|1\rangle$ for a qubit is E_{01}. The *frequency* of this qubit f_{01} is defined as $f_{01} = E_{01}/h$, where h is the

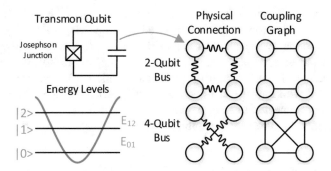

Fig. 2 Superconducting qubit and connection

Planck constant. Similarly, we use f_{12} to represent the energy gap between $|1\rangle$ and $|2\rangle$. For a typical qubit design with effective operations [21], f_{01} and f_{12} are about 5 GHz and 4.66 GHz, respectively. The anharmonicity of this qubit is defined to be $\delta = f_{12} - f_{01}$, which is -340 MHz under this typical design [37, 43].

2.2.2 Qubit Layout

The superconducting physical qubits are confined on a two-dimensional planar substrate. Although the qubit placement can be flexible, major vendors fabricate the qubits in a regularized structure to ensure scalability and reduce the fabrication complexity. For example, IBM's 16-qubit and 20-qubit chips [44] placed their qubits on the nodes of 2×8 and 4×5 lattices, respectively. Google's 72-qubit chip placed its qubits on some nodes of an 11×12 lattice [45].

2.2.3 Qubit Connection

To enable two-qubit gates between two physical qubits, resonators, also known as qubit buses, are employed to connect nearby qubits [42]. For example, Fig. 2 shows two types of commonly used buses. The first one is a 2-qubit bus connecting two physical qubits. The second one is a 4-qubit bus, which connects four physical qubits in a square together. The coupling graphs of these two types of buses are shown on the right. Compared with a 2-qubit bus, 4-qubit bus supports two-qubit gates on not only the four-qubit pairs on the edges but also two-qubit pairs on the diagonals.

2.2.4 Qubit Mapping

It is usually assumed that a two-qubit gate can be applied on arbitrary two logical qubits in a quantum program but some two-qubit gates may not be executable due to the limited qubit connection on a superconducting quantum processor. On the hardware side, this problem can be relieved by employing more physical qubit connections so that two-qubit gates can be directly supported on more qubit pairs. On the software side, a qubit-remapping compiler [46] can resolve the dependency of the remaining unexecutable two-qubit gates while additional operations must be introduced with longer execution time and higher error rate. Therefore, more physical qubit connections can help with the overall performance by allowing native two-qubit gates on more physical qubit pairs.

2.2.5 Fabrication Variation

Variation is inevitable when fabricating a superconducting quantum processor. If a qubit is designed to have frequency f, the actual frequency after fabrication will be $f' = f + n_f$, where n_f satisfies Gaussian distribution $N(0, \sigma)$. σ is the fabrication precision parameter, which is around $130\,\text{MHz} \sim 150\,\text{MHz}$ under IBM's state-of-the-art fabrication technology [36] and can be further calibrated to $14\,\text{MHz}$ using laser-annealing technology [47]. Such noise makes it hard to predict the post-fabrication frequency precisely, which brings the probability of frequency collision.

2.2.6 Frequency Collision

When two or three qubits are connected, *frequency collision* may happen and cause defects on the device. Figure 3 summaries seven qubit frequency collision conditions in IBM's devices [20, 36]. On the left is a table showing the conditions and thresholds of different collision situations. Conditions 1, 2, 3, and 4 involve two connected qubits (j and k). Conditions 5, 6, and 7 involve three qubits of which two qubits (k and i) both connect to the other qubit j. The approximate equations and the corresponding thresholds determine whether one frequency collision happens. For example, if qubit j and k are connected and $|f_j - f_k| < 17\,\text{MHz}$, then the first condition is satisfied and frequency collision occurs. Note that the fourth condition

	Conditions	Thresholds
1	$f_j \cong f_k$	$\pm 17 MHz$
2	$f_j \cong f_k - \delta/2$	$\pm 4 MHz$
3	$f_j \cong f_k - \delta$	$\pm 25 MHz$
4	$f_j > f_k - \delta$	
5	$f_i \cong f_k$	$\pm 17 MHz$
6	$f_i \cong f_k - \delta$	$\pm 25 MHz$
7	$2f_j + \delta \cong f_k + f_i$	$\pm 17 MHz$

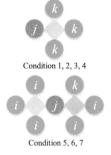

Fig. 3 Frequency collision conditions [20, 36]

has no threshold because it is an inequality rather than an approximate equation. On the right is a graphical illustration, showing the geometric locations of the qubits that may have frequency collisions of different conditions in two subfigures. Each circle represents one qubit and the gray square represents a 4-qubit bus connecting the four surrounding qubits.

3 Quantum Program Profiling

The first step towards the development of an architecture-specific quantum processor for both high performance and yield rate is to determine what program information we should focus on. There are several different types of components in a quantum circuit but not all of them will significantly affect the hardware design. Our target program component(s) should satisfy two conditions: (1) the component's execution is a performance bottleneck which can be dramatically improved with optimized hardware support, and (2) the component's required hardware should significantly affect the yield rate.

We found that two-qubit gates can be a key factor to bridge performance and yield. To execute two-qubit gates on a quantum processor with limited qubit-to-qubit coupling, a large number of additional operations are introduced to satisfy their dependencies. But implementing two-qubit gates on two physical qubits requires on-chip qubit connections which can lower the yield rate through increasing the probability of frequency collision. Therefore, we give logical qubits and qubit pairs priorities based on the number of two-qubit gates involved to help with the following architecture design. Critical qubits and qubit pairs will have more hardware support to improve the efficiency of the generated architectures.

These remaining components, single-qubit gates, initialization, and measurement operations, do not involve qubit-to-qubit interactions and all happen locally on individual qubits when they are implemented on hardware. As a result, hardware support for these components will not affect the chip yield through frequency collision.

3.1 Profiling Method

As discussed above, our profiling will focus on the logical qubits and the two-qubit gates. Figure 4 shows an example to illustrate the profiling procedure. Suppose we have a quantum circuit as shown in Fig. 4a. It has 5 logical qubits denoted by $q_{0,1,2,3,4}$. All of them are initialized to be $|0\rangle$. Then some single-qubit gates and two-qubit gates are applied. Measurement operations are at the end.

We first ignore all single-qubit gates, initialization, and measurement operations. Then we create a logical coupling graph, in which each vertex represents one logical qubit in the circuit. Two vertices are connected by an undirected edge if there exists

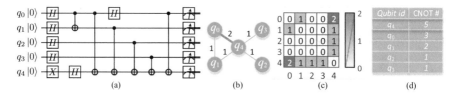

Fig. 4 Example of the profiling method, (**a**) input circuit, (**b**) logical coupling graph, (**c**) coupling strength matrix, (**d**) coupling degree list

two-qubit gates applied on the two corresponding logical qubits. The weight of an edge is the number of two-qubit gate instances on the two connected vertices. In this example, Fig. 4b shows the generated graph for the example circuit. The weight of the edge between vertex q_0 and vertex q_4 is 2 since there are two two-qubit gates on q_0 and q_4. For all other edges, the weight is 1 because there is only one two-qubit gate on each of those qubit pairs. The first profiling result is the weighted adjacency matrix of the logical coupling graph, namely the *coupling strength matrix*. The element with indices (i, j) represents the number of two-qubit gates between q_i and q_j. Figure 4c shows the *coupling strength matrix* for the example circuit. Note that *coupling strength matrix* is always a symmetric matrix.

The second result is *coupling degree list*. For each qubit, we sum the weights of edges that connect to its corresponding vertex and define the number of two-qubit gates applied on it as the *coupling degree* of one qubit. If one qubit is associated with more two-qubit gates in a quantum circuit than other qubits, this qubit will use the physical qubit connections more frequently when executing on the chip. Naturally, we should pay more attention to those qubits with larger coupling degree. Therefore, all qubits are placed in a sorted list, namely the *coupling degree list*. Figure 4d is the *coupling degree list* in this example. The first one in this list is q_4 because it has the largest coupling degree. All qubits are in a descending order.

3.2 Gate Pattern Examples

In this section, we show the existence of distinct two-qubit gate patterns and discuss the opportunity for application-specific architecture design with two examples. Figure 5 shows their *coupling strength matrices*. On the left is an 8-qubit UCCSD ansatz for VQE, a quantum simulation algorithm [4]. The high coupling strength qubit pairs form a chain structure marked by a red rectangle. Q_0 and Q_1 have a large number of two-qubit gates between them, as well as $\{Q_1Q_2, Q_2Q_3, \cdots, Q_6Q_7\}$. For other qubit pairs, the coupling strength is much lower (only about 10%). On the right is a 15-qubit quantum arithmetic function [48]. The coupling strength among $Q_0Q_1 \cdots Q_5$ are 0 since there are no two-qubit gates on any two of them. However, there is a large number of two-qubit gates where one qubit is in the set $Q_{7,8,9,10}$ and the other qubit is in the set $Q_{10,11,12}$ (marked by a red circle). The analysis of these two motivating examples provides us two observations:

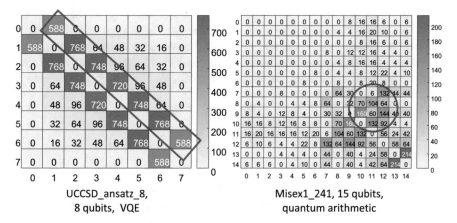

Fig. 5 Qubit coupling strength pattern examples

1. The numbers of two-qubit gates on different logical qubit pairs can vary dramatically in a real quantum program.
2. Different types of quantum programs can have different two-qubit gate patterns.

These observations suggest that quantum processors can be customized for different programs with different patterns. An efficient architecture can focus on supporting the high-density coupling in a quantum program to reduce the number of connections on-chip. For example, a quantum processor with an 8-qubit chain structure (8 qubits and 7 qubit connections) can immediately support most of the two-qubit gates in the 8-qubit UCCSD ansatz program. The rest of the two-qubit gates can be supported through remapping without introducing too many additional operations because the total number of the remaining two-qubit gates is relatively small. Such application-specific QC accelerators with simplified architectures can be a more realistic goal in the near term than a general-purpose quantum processor with a large number of hardware resources.

4 Architecture Design

After a quantum circuit is profiled, a straightforward quantum processor architecture for such a circuit is to organize the on-chip qubits and qubit connections directly based on the logical coupling graph. However, we must consider the physical constraints for a practical architecture. For example, a logical coupling graph may not be perfectly fabricated on hardware since the allowed connections among superconducting qubits are very limited. Moreover, we hope to improve the yield rate by delivering architecture designs with fewer hardware resources. Therefore, the proposed hardware design flow must not only invest more hardware resource on frequent operations based on the profiling results, but must also obey the physical constraints on the hardware components arrangement.

To accomplish such a complicated task in a scalable way, we decouple the hardware design procedure into three subroutines and each subroutine focuses on different architecture components, i.e., qubit layout, connection, and frequency. For each subroutine, we first review the difficulty and the physical constraints considered. Then we discuss the design objectives, and how they are achieved in the proposed design algorithms.

4.1 Layout Design

The first step is to determine where to place the qubits. To ensure scalability and modularity, we follow the convention from major vendors introduced in Sect. 2 and will only place qubits on the nodes of a 2D lattice. We start from a large 2D lattice, in which each node is initialized to be empty (Fig. 6a). Then physical qubits can be placed in the empty nodes and one node can contain at most one qubit.

There are many ways to place a given number of qubits on a 2D lattice. For example, 16 qubits can constitute a 4×4 lattice, a 2×8 lattice, or other more irregular structures. But we need to select one-qubit layout that is most suitable for executing the program, i.e., most operations can be directly supported or indirectly supported with low overhead. The objectives of this qubit layout design subroutine are summarized as follows.

- Since we need to consider the profiling information, we create a pseudo mapping between logical qubits in the profiled program and the physical qubits in hardware architecture to be delivered. For two logical qubits with a large number of two-qubit gates between them, we hope to place their corresponding physical qubits in adjacent nodes so that later those two-qubit gates can be directly supported by the connection between the two physical qubits.
- One physical qubit can only have a limited number of directly connected qubits. For those two-qubit gates that cannot be directly supported, we hope to reduce the amount of additional operations introduced for remapping the qubits.

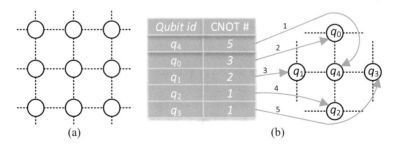

Fig. 6 (**a**) Empty lattice, (**b**) qubit placement example

---

Algorithm 1: Qubit Placement on 2D Lattice

Input: *coupling degree list L, coupling strength matrix M*
Output: Geometric coordinates of placed qubits
1 Place the qubit with the largest coupling degree in L at one node with coordinate $(0, 0)$;
2 $R =$ all the qubits remaining; // R is the set of qubits that has not been placed yet.
3 **while** R *is not empty* **do**
      /* Find the next qubit to place   */
4      $qubit\_candidate\_list = \varnothing$;
5   **for** q *in R* **do**
6   **if** q *is connected to any placed qubits* **then**
7              $qubit\_candidate\_list.append(q)$;
8   **end**
9   **end**
10   Find the qubit q with the largest coupling degree in $qubit\_candidate\_list$;
11     $node\_cost = [\ ]$;
      /* Determine the placement location   */
12   **for** *location* of the nodes that are empty and connected to at least one occupied node **do**
        /* Heuristic Cost function   */
13         $node\_cost\,[\textbf{\textit{location}}] =$
            $\sum\limits_{q' \in q.neighbors} M\,[q, q'] * distance\,[\textbf{\textit{location}}, q'.node]$
14   **end**
      /* q' must be placed neighbor qubits   */
15   Place q in the *location* with the minimal score;
16     $R.remove(q)$;
17 **end**

We propose a *coupling-based* qubit placement algorithm to determine the geometric locations of the qubits on a 2D lattice (pseudocode shown in Algorithm 1). We illustrate the algorithm with an example in Fig. 6. First, we put the first qubit in the *coupling degree list*, q_4, on one node of the 2D lattice. Since the initial 2D lattice is empty, the location of q_4 does not matter. We set the geometric coordinate of the first qubit to be $(0, 0)$ and then place the rest qubits around q_4. q_4 has four neighbors, $q_{\{0,1,2,3\}}$, in the logical coupling graph. We need to select the next one to place. By checking the *coupling degree list*, we can see that q_0 is the one with the largest coupling degree. The node occupied by q_4 has four equivalent adjacent nodes and we can place q_0 on any of them. In this example, we select the node to the north of q_4 with coordinate $(0, 1)$. Such an algorithm design ensures that the strongly coupled qubit pairs are given higher priority and placed on adjacent nodes, accomplishing the first objective mentioned above.

Then we need to place q_1 since its coupling degree is larger than that of q_2 and q_3. q_1 is connected to both q_4 and q_0 so that we need a more sophisticated way to evaluate all potential nodes for q_1. We use the function in line 13 of Algorithm 1 to find the node that can make q_1 close to its strong coupled neighbors in the logical coupling graph. This function is the summation over all q_1's placed neighbors. Each term in the summation is the product of the coupling strength between q_1 and one logical coupling neighbor q' and the Manhattan distance between the evaluated node location and the location of q'. After evaluating all the empty nodes that are adjacent to placed nodes q_4 and q_0, we will find that the nodes on the east and west of q_4 are the best ones because they are closest to q_4 but not far away from q_0. Here we select the one on the west of q_4 with coordinate $(-1, 0)$. This summation function can help reduce the number of operations for later remapping and achieve the second design objective.

The remaining qubits can be placed in a similar procedure until all the qubits have been placed on the 2D lattice. In this example, q_2 and q_3 are placed on the nodes with coordinates $(0, -1)$ and $(1, 0)$, respectively. All the qubits have their locations (coordinates) on a 2D lattice where we can fabricate one physical qubit on each occupied node. Finally, the nodes with no qubits are removed.

4.2   Bus Selection

In the second step, we need to connect the placed physical qubits to enable two-qubit gates. The difficulty comes from the large size of the design space. For N qubits, there are $\binom{N}{2}$ distinct qubit pairs. Any of them can be either connected or disconnected so that there are $2^{\binom{N}{2}}$ different cases. Even after considering the nearest neighbor coupling constraint in which one qubit can only connect with few qubits around it on the lattice, the size of the design space is still $O(exp(N))$. More importantly, more qubit connections will improve the performance but lower the yield rate in general so that we need to identify those connections with the most potential performance benefit in a very large design space.

We simplify the connection design problem by considering two types of common buses, 2-qubit bus and 4-qubit bus (shown in Fig. 2). These two types of buses naturally fit in the 2D lattice qubit layout and can be easily fabricated because at most 4 nearby qubits are connected by one bus. After placing the qubits on a 2D lattice in the first step, 2-qubit buses can be directly generated on the edges that connect two occupied nodes but the qubits on a diagonal of a 4-qubit square can never be connected with only 2-qubit buses. Replacing some 2-qubit buses with 4-qubit buses could provide more qubit connection by trading in yield rate while it is not yet clear where to apply the 4-qubit buses can achieve the Pareto-optimal results. The bus selection subroutine was proposed to identify the locations for 4-qubit buses. Other potential bus designs are left as future research directions and will be discussed in Sect. 6.

Instead of considering the nodes in a 2D lattice, we consider the squares that are naturally formed by the edges in the 2D lattice. Each square can be configured to 2-qubit bus or 4-qubit bus. Now the problem is on which squares we should use 4-qubit buses. The size of search space, even for this 4-qubit bus square selection problem, is still $O(exp(N))$. But the simplification allows us to design high-quality heuristics

Fig. 7 (**a**) Prohibited
condition, (**b**) corner case, (**c**)
filtered weight

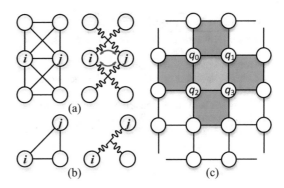

to guide the selection. Before introducing our solution, one additional prohibited condition must be considered.

4.2.1 Prohibited Condition

One physical constraint that we must consider when applying 4-qubit buses is that we cannot have 4-qubit buses in two adjacent squares. The reason is explained with the example in Fig. 7a. Suppose we have two adjacent squares and both of them are using 4-qubit buses. Then there will be two physical connections between qubits i and j. When we use one of the connections, the other one will bring unexpected effects so that employing 4-qubit bus in one square will immediately block using 4-qubit buses in any of its adjacent squares.

Considering the physical constraints mentioned above, the objectives of this step are summarized as follows:

- Since adding more qubit connections will increase the probability of frequency collision and lower the yield, we hope to apply 4-qubit buses on those squares that can benefit the performance most. In other words, the additional connections are expected to directly support as many two-qubit gates as possible.
- Applying 4-qubit bus in one square will block adjacent squares, making it impossible to directly support some two-qubit gates in those blocked squares. This effect should also be considered when selecting the 4-qubit squares.

We propose a 4-qubit bus selection algorithm to select some squares for 4-qubit buses (pseudocode shown in Algorithm 2). In each iteration, one square that could benefit most from a 4-qubit bus will be selected. Users can specify the maximum number of 4-qubit buses they hope to have. By varying the number of selected squares, a series of architectures can be generated with a trade-off between yield and performance.

Algorithm 2: 4-qubit Bus Selection

Input: Geometric coordinates of placed qubits, *coupling strength matrix*, Maximum
number of 4-qubit buses K
Output: Locations of 4-qubit Buses

1 Calculate the cross coupling weight for each square;
2 **while** $K > 0$ **do**
    // Select one square in each iteration
3   **for** *square(i, j) in all squares* **do**
4   *filtered_weight(i, j) = weight(i, j) - weight(i+1, j) - weight(i, j+1) - weight(i-1, j) -*
      weight(i, j-1);
5   **end**
6   **if** *no square available for 4-qubit bus* **then**
7 | Break;
8   **end**
9   Select the square with the highest $filtered\_weight$;
10 Set the weights of squares *(i+1, j), (i, j+1), (i-1, j)*, and *(i, j-1)* to be 0 and mark them
    to be blocked;
11 $K = K - 1$;
12 **end**

To find the most fitting square, we first need to calculate how much one square could benefit from a 4-qubit bus. Since the difference between a 2-qubit bus square and a 4-qubit bus square is whether the qubit pairs on the diagonals are connected, we define the cross-coupling weight for each square as the sum of the coupling strength of the qubit pairs on the diagonals. For the example in Fig. 7c, the cross-coupling weight of the green square is the coupling strength of (q_0, q_3) plus that of (q_1, q_2). A corner case in the coupling weight computation is the square with only 3 qubits (shown in Fig. 7b). In such squares, 4-qubit buses can naturally reduce to 3-qubit buses which support coupling between any two of the three connected qubits. The weight of a 3-qubit square is only the weight of logical coupling between the two qubits on one diagonal since the other diagonal only has one qubit. For example, the weight of the 3-qubit square in Fig. 7b is the (i, j) element in the *coupling strength matrix*. Except for this small modification, 3-qubit squares are treated equally as other 4-qubit squares in our bus selection step. This cross-coupling weight can estimate the potential benefit of applying 4-qubit bus in one square and realize the first objective.

However, the cross-coupling weight is not accurate enough to evaluate the benefit of 4-qubit for a square because the prohibited condition is not yet considered. We design a filter to apply this constraint. For each square, the filtered weight is its original cross-coupling weight minus all its neighbors' weights. For example, in Fig. 7c, the filtered weight of the green square is its original weight minus the weights of the four blue squares. This filter can take the prohibited condition into consideration and achieve the second objective.

After applying the filter, we will select one square with the highest filtered weight. Then we will label the selected square and its adjacent neighbors so that it will no longer be available for future 4-qubit buses. We also change their weights to zero because they should not affect the 4-qubit selection among the remaining squares. The algorithm will iterate again to select the next square until there are no more squares available or we have already applied enough number of 4-qubit buses.

4.3 Frequency Allocation

After the two steps above, we now have a complete coupling topology design of a superconducting quantum processor. In the third step, we need to designate the pre-fabrication frequency of each qubit. A regular frequency designation [36] will assign several different frequencies to the qubits in a repeated pattern. However, the generated qubit layout and connection in our design flow can be irregular since more hardware sources are invested in locations that can benefit the performance most. Thus, we need a more flexible frequency allocation scheme to leverage this unbalanced qubit layout and connection. The objective of this step is to minimize the probability of post-fabrication frequency collision and improve the yield rate. The physical constraints are the frequency collision conditions in Fig. 3.

Finding the qubit frequency allocation plan to maximize the yield rate is a hard problem. The complex collision conditions make it difficult to find an analytic expression for the yield rate and a brute-force search over all possible frequency configurations will be very time-consuming. For example, if there are M candidate frequencies for each qubit and we have N qubits in total, the total number of possible frequency configurations is M^N. For each of these potential configurations, we need to run a yield simulation (introduced in Sect. 4.3.1) and then select the one with maximal yield rate. This method is not acceptable due to its high complexity. We propose to optimize the qubit frequency allocation algorithm based on the facts that (1) the physical qubits in the geometric center of the qubit lattice are more likely to be involved in a frequency collision since they usually have more qubit connections, and (2) frequency collision only happens among nearby qubits.

Algorithm 3: Frequency Allocation

Input: Qubit Location and Connection
Output: Frequency Configuration of Each Qubit
1 Select the qubit in the geometric center of the placed qubits and set its frequency to be the middle of the allowed frequency range;
2 **repeat**
3   Find the next qubit q_i in breadth-first traversal order;
4   **for** $temp\_freq$ in all frequency samples **do**
5   Set the frequency of q_i to be $temp\_freq$;
6   Simulate the yield rate within q_i's local region;
7   **end**
8   Assign the frequency with maximal yield rate to q_i;
9 **until** *the frequencies of all qubits are determined*;

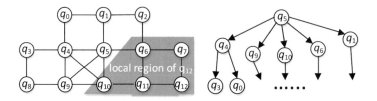

Fig. 8 Breath first frequency allocation

Our algorithm determines the qubit frequencies from the center to the periphery (pseudocode shown in Algorithm 3). Since this step is purely about hardware, the input of our algorithm is only the qubit location and connection generated from the previous two subroutines. To reduce the manufacturing difficulty and help prevent the collision condition 4, we follow the convention from IBM and set an allowed frequency interval 5.00 GHz to 5.34 GHz. All pre-fabrication frequencies are limited within this interval. First, we locate the qubit that is closest to the center of the qubit lattice and assign its frequency to be the center of the allowed frequency interval. Then we apply breadth-first traversal on the coupling graph from the first qubit in the center. For example, q_5 is the center qubit in the example shown in Fig. 8. In the breadth-first traversal, we will first access $q_{4,9,10,6,1}$ as shown on the right. Each time we access one new qubit, we will immediately determine its frequency. A list of candidate frequencies is prepared. In this study, the candidate frequencies are $5.00, 5.01, 5.02, \ldots, 5.33, 5.34$ GHz to achieve an accuracy of 0.01 GHz. We can also have more candidate frequencies but it will take more time to evaluate all of them.

To evaluate a candidate frequency on a new qubit, we temporarily assign the candidate frequency to the new qubit and then simulate the yield rate within its local region. The local region of a qubit is defined as a sub-graph of the original chip coupling graph in which a qubit may collide with the new qubit. For example, in Fig. 8, when we are searching for the best frequency of q_{12}, the local region is marked in blue. Note that it is necessary to consider two hops when allocating frequency for one qubit because the frequency collision conditions in row 5, 6, and 7 of Fig. 3 involve 3 connected physical qubits. Qubits not in this region like q_5 cannot collide with q_{12}. We will select the frequency with the maximal yield rate and assign it to the new qubit. Now the time complexity of the frequency allocation algorithm is $O(MN)$ where M is the number of candidate frequencies and N is the number of qubits.

4.3.1 Yield Simulation

We developed a yield simulator based on IBM's yield model [20, 36]. The fabrication process can be modeled by adding a Gaussian noise $N(0, \sigma)$ to the pre-fabrication frequency of a qubit to generate its post-fabrication frequency where

σ is the fabrication precision parameter. For a given superconducting quantum processor design, we estimate its yield rate through Monte Carlo simulation. Each time we will simulate if one fabrication is successful. We first generate the post-fabrication frequencies by adding a random noise sampled from Gaussian distribution mentioned above. Then we check if any frequency collision condition listed in Fig. 3 occurs in the post-fabrication frequencies. If so, this fabrication fails. Otherwise, it is successful. All possible cases are taken into account. For example, we will examine the two frequencies of all connected physical qubit pairs for condition 1, 2, 3, and 4. If they meet any one of the inequalities of the conditions, frequency collision is considered to occur in this simulation. This simulation process is repeated many times. The yield rate can be estimated by the ratio between the number of successful simulations and the total number of simulations.

5 Evaluation

To demonstrate that the proposed application-specific architecture design flow can deliver hardware designs with better Pareto-optimal results in terms of performance and yield rate, we conduct experiments over various benchmarks to show not only the overall improvement but also the breakdown of benefits from each of our hardware design subroutines.

5.1 Experiment Setup

5.1.1 Benchmarks

Twelve quantum programs are collected from IBM's Qiskit [49] and RevLib [48], or compiled from ScaffCC [28]. These benchmarks cover several important domains (e.g., simulation, arithmetic) and have various sizes (from 7- to 16-qubit) for a versatility test of the proposed design flow.

5.1.2 Metrics

To evaluate the efficiency of an architecture, we need both the yield rate and performance. An architecture with a higher yield rate can be successfully fabricated with fewer attempts, indicating a lower hardware cost. In our experiments, the yield rate is simulated with IBM's yield model [20, 36] as introduced in Sect. 4.3.1. For the performance evaluation, we adopt the total post-mapping gate count metric widely used in previous studies [16–18]. More gates lead to longer execution time and a larger probability of error on QC devices. If a hardware architecture could execute the program with fewer gates, then its performance is considered to be better.

5.1.3 Yield Simulation Configuration

The number of trials in the Monte Carlo simulation for each architecture is 10,000~ 100,000, which is 10 ~ 100× of that used in IBM's experiments [20, 37, 50] to ensure the simulation accuracy. The fabrication precision parameter σ is set to be 30 MHz, a realistic extrapolation of progress in hardware by IBM [36, 37]. IBM has improved the σ from 200 MHz [51] to 130 MHz [36] in the last few years and 30 MHz is a reasonable projection to achieve a useful yield as predicted by IBM [37].

5.2 Experiment Methodology

To illustrate the benefit of our design flow, five experiment configurations are designed to show the overall improvement and the performance/yield trade-off gain at each of the three subroutines in Sect. 4. Among them, **gp** is a set of general-purpose architectures not tailored for any applications. The remaining four configurations are application-specific architectures generated by the entire or part of the proposed design flow.

gp We use 4 regular design schemes as the baseline configuration for general-purpose architectures. It has two layout options, a 2×8 lattice with 16 qubits, and a 4×5 lattice with 20 qubits. The qubit connection design can be either 2-qubit bus only or using 4-qubit buses as many as possible. In total, there are four architectures combining the layout and connection options and they are labeled by (1)–(4) in Fig. 9. The frequency allocation scheme is a 5-frequency scheme [36, 37]. The five frequencies are an arithmetic progression from 5 GHz to 5.27 GHz and their arrangement is also in Fig. 9.

eff-full We apply all three subroutines and generate a series of efficient superconducting quantum processor architectures by varying the number of 4-qubit buses. The number of designs we can obtain for a quantum program depends on the

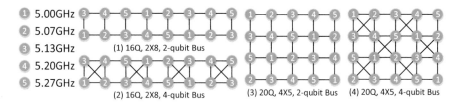

Fig. 9 Baseline qubit frequency, layout, and connection designs

number of qubits as more qubits can provide more squares to apply 4-qubit buses in the generated layout. In this study, we obtain the **eff-full** data series through iterating over all possible numbers of 4-qubit buses in the second subroutine for bus selection. This experiment can show the overall architecture design improvement when comparing with the baseline **gp**.

eff-5-freq We only apply the first two subroutines to generate qubit layout and connection design but the frequency allocation is done with the 5-frequency scheme in the baseline **gp**. The yield benefit from the proposed frequency allocation algorithm can be demonstrated by comparing with results from **eff-full**.

eff-rd-bus We keep the first and the third subroutines but randomly select some squares to employ 4-qubit buses with the prohibited condition constraint satisfied. This will demonstrate the effect of our filtered-weight-based 4-qubit bus selection algorithm by comparing with results from **eff-full**.

eff-layout-only We apply our profiling method and perform a layout design. The connection design has two options. One is only using 2-qubit buses. The other is using 4-qubit buses as much as possible. The frequency design follows the baseline **gp**. The benefit of our layout optimization can be shown when comparing with the results from **gp**.

For each benchmark, we run all the five configurations to generate different superconducting quantum processor architectures with different yield rates. Then we apply one state-of-the-art qubit mapping algorithm [18] on these architectures to obtain the total number of gates when running the generated or baseline architectures.

5.3 Overall Improvement

Figure 10 shows the result of yield and performance for all benchmarks and the five experiment configurations. There are 12 subfigures and one subfigure contains the results of the five experiment configurations for one benchmark. The X-axis represents the normalized inversed post-mapping gate count. Data points on the **right** represent fewer post-mapping gates and have better performance. The Y-axis represents the yield rate and data points on the **top** have higher yield rates. The legend at the bottom of Fig. 10 shows the markers for the five configurations. The data points for the four designs in the baseline are labeled by (1), (2), (3), and (4), according to Fig. 9.

5.3.1 Optimality

The optimal solution in our evaluation means the Pareto-optimal solution in terms of post-mapping gate count and yield rate. A series of architectures with better Pareto-optimal results can be generated by our design flow as the data of **eff-full** is on the upper right of **gp**. The most simplified designs (the most left top blue triangle

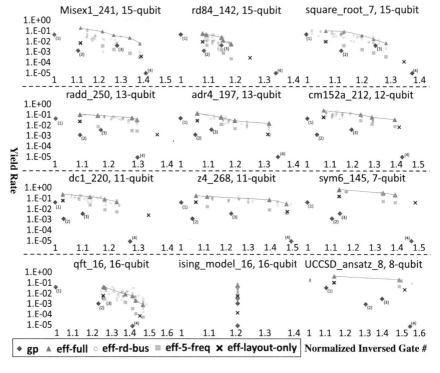

Fig. 10 Yield v.s. normalized reciprocal of post-mapping gate count

data point in **eff-full**, zero 4-qubit buses) generated by our design flow outperforms the 16-qubit baseline design (data point (1) in **gp**) without 4-qubit buses in both performance ($\sim$7.7%) and yield rate ($\sim$4$\times$). Compared with the 16-qubit baseline with four 4-qubit buses (data point (2) in **gp**), our designs with zero 4-qubit buses achieve over 100$\times$ better yield rate with <1% performance loss. On the other side, compared with the 20-qubit chip design with six 4-qubit buses (the baseline design with the most hardware resources, data point (4) in **gp**), the designs with the maximum number of 4-qubit buses generated from our design flow (the data points on the most bottom right in **eff-full**) have over 1000$\times$ yield rate improvement on average with only about 3.5% performance loss.

5.3.2 Controllability

The proposed design flow can easily control the trade-off between yield and performance by only changing the number of 4-qubit buses without traversing across, or sampling a large number of designs in, the entire search space. Depending on the number of qubits in different target programs, we can trade in around 10$\times$ $\sim$ 50$\times$ yield rate for 10% $\sim$ 33% performance improvement.

5.3.3 Special Case

The results of *ising_model* are significantly different because the logical qubit coupling in this benchmark forms a chain structure. The mapping algorithm can always find the perfect initial mapping without inserting additional operations. As a result, the post-mapping gate count is the same for all tested hardware architectures. All data points for this program lie in one vertical line. Only one architecture is generated from our design flow because there is no need to add 4-qubit bus. All the two-qubit gates can be executed through the edges on the 2D lattice. There are no two-qubit gates applied on two qubits on a diagonal because of the chain coupling structure. In this case, 4-qubit buses can only lower the yield rate without improving the performance.

5.4 Effects from Individual Subroutines

The overall improvement has already been discussed, but one interesting question is how much improvement the layout and connection optimization contribute and how much comes from the optimized yield allocation directly. The five configurations decouple the proposed design flow and provide a breakdown of the effect of individual subroutines.

5.4.1 Effect of Layout Design

The difference between **gp** and **eff-layout-only** illustrates the effect of layout design since the rest two subroutines are the same. An architecture with more hardware resources is expected to provide higher performance by allowing more flexibility in qubit mapping. But our optimized layout design could use comparable or fewer hardware resources while the performance can be even better. For example, we compare the 2-qubit bus only data point (the upper left one) with the 16-qubit baseline with four 4-qubit buses (labeled by (2) in each subfigure). **eff-layout-only** provides better or comparable performance most of the time with about 35× yield improvement on average. The improvement at this step depends on the program size and programs with fewer qubits will use fewer qubits and connections in an optimized architecture. This result proves that our layout design could generate qubit layout with high performance but using much fewer hardware resource for different programs.

5.4.2 4-Qubit Bus Selection Quality

By comparing the results from **eff-full** and **eff-rd-bus**, we can see that the architectures generated from our bus selection algorithm are better than that of random selection in trading in yield for performance most of the time. The data points of **eff-rd-bus** reveal the distribution of the yield and performance sampled from random bus designs. Note that the performance of **eff-rd-bus** is usually confined by the two data points in **eff-layout-only** because adding connections can improve the performance most of the time. For most benchmarks except *qft*, the results from **eff-full** are close to the upper bound formulated by the random samples, which shows that our weight-based bus selection could generate a series of near Pareto-optimal hardware architectures with various numbers of qubit connections.

The result of *qft* is much worse than that of other programs due to the unique uniform two-qubit gate pattern in this program. The number of two-qubit gates between arbitrary two logical qubits is always two in *qft*. Therefore, all non-diagonal elements in the *coupling strength matrix* of the *qft* program are two. Then in bus selection subroutine, all the squares share the same weight and the weight-based selection is the same as random selection.

For the two small benchmarks, *sym6* and *UCCSD_ansatz*, the number of available squares in the generated qubit layout is small and there are very few options when applying 4-qubit buses. Therefore, most of the architectures generated from the random 4-qubit bus selection are the same as those from the proposed design flow, which makes the results from **eff-full** and **eff-rd-bus** very close.

5.4.3 Frequency Allocation Optimization

By comparing **eff-full** and **eff-5-freq**, we can see that the proposed frequency allocation algorithm provides about $10\times$ yield rate improvement on average. This improvement is slightly worse when the yield from the baseline 5-frequency is already high, e.g., results from *sym6* and *UCCSD_ansatz*. The fabrication variance makes the ideal yield 100% unreachable and it is hard to optimize yield when it is already high.

6   Discussion

In this chapter, we studied application-specific efficient superconducting quantum processor design. In particular, we formalize the architecture design for superconducting quantum processors with three key steps, each of which comes with an optimization subroutine. This is the first attempt, to the best of our knowledge, to identify the optimization opportunity from the architecture level to push forward the balance between QC performance and hardware yield rate. Effort towards this direction can be of significant demand in the near-term QC with limited computational resource and immature fabrication technology.

Although we show that improved Pareto-optimal designs can be generated with a static program analysis and three optimized design algorithms, several future research directions can be explored as with any initial research.

6.1 Improving Profiling Method

This study focused on the logical qubit coupling topology in a quantum program but other patterns may also be leveraged. We omitted the temporal information of the two-qubit gates and all information about other program components. But the locations of two-qubit gates in a quantum program may also be leveraged for finer-grained evaluation of the coupling strength for different logical qubit pairs at different times during the execution. The single-qubit patterns can also help with the basic gate set design.

6.2 Exploring More Design Space

In the proposed design flow, the number of physical qubits is the same as that of logical qubits for higher yield rate. However, we can still add auxiliary physical qubits since they can also be used during the qubit routing, trading in more yield rate for higher performance. How to add auxiliary qubit to appropriate locations and how to connect them are interesting problems to explore in the future. To ensure modularity and scalability, the qubits are forced to be embedded in a 2D lattice and only consider two types of buses lying in the lattice. However, the qubit placement and connection could be more flexible if we trade in part of the scalability. For example, one bus could also connect more than four qubits [52]. The design space in this direction is not yet explored.

6.3 Optimizing Frequency Allocation

This study tried to optimize the qubit frequency selection from the center to periphery and only searched for the optimal frequency for one qubit, resulting in a sub-optimal frequency allocation. A global optimization like formal methods can be explored to further optimize the frequency allocation result. One alternative approach to resolve the frequency collision issue is to use flux-tunable transmon qubits [22], of which the frequencies can be dynamically tuned with additional control signals. The design trade-off of different types of qubits is not yet explored and additional signals bring more noise and increase the control complexity. The proposed design flow is still valuable even with frequency-tunable qubits because the simplified architectures with fewer the on-chip connections can not only reduce the fabrication complexity but also benefit the overall performance by lowering the crosstalk error.

7   Related Work

This study ranges across multiple topics, i.e., program profiling, superconducting processor design, application-specific design, qubit mapping. We briefly introduce related work for all of them.

7.1   Application-specific Design

The closest related work is SPARQS, a superconducting planar architecture proposed by Wilhelm et al. [34, 35] targeting a specific Fermi-Hubbard model simulation program. However, they only provide an implementation-independent design from theoretical physics level. This study formalizes a systematic end-to-end design flow with automatic program profiling and realistic physical constraints included, for the first time. With no limitation on the target program, we can generate a series of Pareto-optimal hardware architecture designs in a controllable way.

7.2   Quantum Program Profiling and Analysis

Program profiling and analysis are very important for software and compiler optimization. Previous works on quantum program analysis [28–33] have studied entanglement, termination, non-cloning checking, etc. The profiling method in this study is proposed to guide the hardware design, fulfilling a different goal.

7.3   Superconducting Quantum Processors

As one of the most promising candidate technologies to implement QC, superconducting quantum techniques have been employed in two mainstream QC computational models. The circuit model based processors [44, 45, 53] support quantum circuit model [40] and the quantum annealers [54] can implement adiabatic QC [55]. Their programming model and hardware architecture are different for these two QC approaches. The design flow in this study is proposed for circuit model based quantum processors while efficient quantum annealer design can be a future research direction.

7.4 Qubit Mapping

Formal and heuristic methods have been attempted to solve this problem [16–18, 56, 57] and minimize the total gate count. Recently several studies [10, 58, 59] have applied the actual gate error rates for fine-grained optimization. All these optimizations are pure software-level modification. This study attempts to improve the performance by reducing the mapping overhead from the hardware level. We adopt the gate count metric to estimate the mapping overhead since our experiments are performed on artificial hardware architectures.

8 Conclusion

The demand for larger computational capability in a superconducting quantum processor naturally calls for more hardware resources which will also increase the design complexity and lower the yield rate. This study explored application-specific architecture design for superconducting quantum processors to achieve both high performance and higher yield rate. Gate patterns in a quantum program can be extracted by the proposed profiling method and then utilized in the follow-up hardware architecture design. Three subroutines are designed to generate the qubit layout, connection, and frequency, respectively, with physical constraints taken into consideration. Experimental results show that the proposed design flow could deliver architectures with both high yield rate and performance automatically for different applications except those with extremely special gate patterns.

References

1. A.W. Harrow, A. Hassidim, S. Lloyd, Phys. Rev. Lett. **103**(15), 150502 (2009)
2. E. Farhi, J. Goldstone, S. Gutmann, Preprint, arXiv:1411.4028 (2014)
3. S. McArdle, S. Endo, A. Aspuru-Guzik, S. Benjamin, X. Yuan, Preprint arXiv:1808.10402 (2018)
4. A. Peruzzo, J. McClean, P. Shadbolt, M.H. Yung, X.Q. Zhou, P.J. Love, A. Aspuru-Guzik, J.L. O'brien, Nat. Commun. **5**, 4213 (2014)
5. M.H. Devoret, R.J. Schoelkopf, Science **339**(6124), 1169 (2013)
6. H. Paik, D.I. Schuster, L.S. Bishop, G. Kirchmair, G. Catelani, A.P. Sears, B.R. Johnson, M.J. Reagor, L. Frunzio, L.I. Glazman, S.M. Girvin, M.H. Devoret, R.J. Schoelkopf, Phys. Rev. Lett. **107**, 240501 (2011). https://doi.org/10.1103/PhysRevLett.107.240501. https://link.aps.org/doi/10.1103/PhysRevLett.107.240501
7. R. Barends, J. Kelly, A. Megrant, D. Sank, E. Jeffrey, Y. Chen, Y. Yin, B. Chiaro, J. Mutus, C. Neill, P. O'Malley, P. Roushan, J. Wenner, T.C. White, A.N. Cleland, J.M. Martinis, Phys. Rev. Lett. **111**, 080502 (2013). https://doi.org/10.1103/PhysRevLett.111.080502. https://link.aps.org/doi/10.1103/PhysRevLett.111.080502
8. Y. Chen, C. Neill, P. Roushan, N. Leung, M. Fang, R. Barends, J. Kelly, B. Campbell, Z. Chen, B. Chiaro, A. Dunsworth, Phys. Rev. Lett. **113**(22), 220502 (2014)

9. Y. Shi, N. Leung, P. Gokhale, Z. Rossi, D.I. Schuster, H. Hoffmann, F.T. Chong, in *Proceedings of the Twenty-Fourth International Conference on Architectural Support for Programming Languages and Operating Systems* (ACM, New York, 2019), pp. 1031–1044

10. P. Murali, J.M. Baker, A. Javadi-Abhari, F.T. Chong, M. Martonosi, in *Proceedings of the Twenty-Fourth International Conference on Architectural Support for Programming Languages and Operating Systems, ASPLOS 2019*, Providence, April 13–17 (ACM, New York, 2019), pp. 1015–1029

11. X. Fu, M.A. Rol, C.C. Bultink, J. van Someren, N. Khammassi, I. Ashraf, R.F.L. Vermeulen, J.C. de Sterke, W.J. Vlothuizen, R.N. Schouten, C.G. Almudever, L. DiCarlo, K. Bertels, in *Proceedings of the 50th Annual IEEE/ACM International Symposium on Microarchitecture* (IEEE/ACM, Piscataway/New York 2017), pp. 813–825

12. J.P. van Dijk, E. Charbon, F. Sebastiano, Preprint. arXiv:1811.01693 (2018)

13. J. Koch, M.Y. Terri, J. Gambetta, A.A. Houck, D. Schuster, J. Majer, A. Blais, M.H. Devoret, S.M. Girvin, R.J. Schoelkopf, Phys. Rev. A **76**(4), 042319 (2007)

14. D.C. McKay, S. Filipp, A. Mezzacapo, E. Magesan, J.M. Chow, J.M. Gambetta, Phys. Rev. Appl. **6**(6), 064007 (2016)

15. A.W. Cross, L.S. Bishop, S. Sheldon, P.D. Nation, J.M. Gambetta, Preprint. arXiv:1811.12926 (2018)

16. M.Y. Siraichi, V.F.d. Santos, S. Collange, F.M.Q. Pereira, in *Proceedings of the 2018 International Symposium on Code Generation and Optimization* (ACM, New York, 2018), pp. 113–125

17. A. Zulehner, A. Paler, R. Wille, in *Design, Automation & Test in Europe Conference & Exhibition (DATE), 2018* (IEEE, Piscataway, 2018), pp. 1135–1138

18. G. Li, Y. Ding, Y. Xie, in *Proceedings of the Twenty-Fourth International Conference on Architectural Support for Programming Languages and Operating Systems* (ACM, New York, 2019), pp. 1001–1014

19. P. Murali, N.M. Linke, M. Martonosi, A.J. Abhari, N.H. Nguyen, C.H. Alderete, in *Proceedings of the 46th International Symposium on Computer Architecture*, ISCA '19 (ACM, New York, 2019), pp. 527–540. https://doi.org/10.1145/3307650.3322273. http://doi.acm.org/10.1145/3307650.3322273

20. M. Brink, J.M. Chow, J. Hertzberg, E. Magesan, S. Rosenblatt, in *2018 IEEE International Electron Devices Meeting (IEDM)* (IEEE, Piscataway, 2018), pp. 1–6

21. E. Magesan, J.M. Gambetta, Preprint. arXiv:1804.04073 (2018)

22. J. Kelly, R. Barends, A. Fowler, A. Megrant, E. Jeffrey, T. White, D. Sank, J. Mutus, B. Campbell, Y. Chen, Z. Chen, Nature **519**(7541), 66 (2015)

23. S. Rosenblatt, J.S. Orcutt, J.M. Chow, Laser annealing qubits for optimized frequency allocation (2019). US Patent App. 10/340,438

24. T. Chen, Z. Du, N. Sun, J. Wang, C. Wu, Y. Chen, O. Temam, in *ACM SIGPLAN Notices*, vol. 49 (ACM, New York, 2014), pp. 269–284

25. S. Han, X. Liu, H. Mao, J. Pu, A. Pedram, M.A. Horowitz, W.J. Dally, in *2016 ACM/IEEE 43rd Annual International Symposium on Computer Architecture (ISCA)* (IEEE, Piscataway, 2016), pp. 243–254

26. T.J. Ham, L. Wu, N. Sundaram, N. Satish, M. Martonosi, in *2016 49th Annual IEEE/ACM International Symposium on Microarchitecture (MICRO)* (IEEE, Piscataway, 2016), pp. 1–13

27. J. Ahn, S. Hong, S. Yoo, O. Mutlu, K. Choi, ACM SIGARCH Comput. Architect. News **43**(3), 105 (2016)

28. A. JavadiAbhari, S. Patil, D. Kudrow, J. Heckey, A. Lvov, F.T. Chong, M. Martonosi, Parallel Comput. **45**, 2 (2015)

29. M. Ying, Y. Feng, Acta Informatica **47**(4), 221 (2010)

30. M. Ying, N. Yu, Y. Feng, R. Duan, Sci. Comput. Program. **78**(9), 1679 (2013)

31. S. Ying, Y. Feng, N. Yu, M. Ying, in *International Conference on Concurrency Theory* (Springer, Berlin, 2013), pp. 334–348

32. K. Honda, Preprint. arXiv:1511.01572 (2015)

33. S. Perdrix, in *International Static Analysis Symposium* (Springer, Berlin, 2008), pp. 270–282

34. P.L. Dallaire-Demers, F.K. Wilhelm, Phys. Rev. A **94**(6), 062304 (2016)
35. P.J. Liebermann, P.L. Dallaire-Demers, F.K. Wilhelm, Preprint arXiv:1701.07870 (2017)
36. S. Rosenblatt, J. Hertzberg, J. Chavez-Garcia, N. Bronn, H. Paik, M. Sandberg, E. Magesan, J. Smolin, J.B. Yau, V. Adiga, M. Brink, J.M. Chow, Bull. Am. Phys. Soc. (2019)
37. C. Chamberland, G. Zhu, T.J. Yoder, J.B. Hertzberg, A.W. Cross, Preprint, arXiv:1907.09528 (2019)
38. J.B. Hertzberg, R.O. Topaloglu. Quantum circuit topology selection based on frequency collisions between qubits (2020). US Patent App. 16/449,976
39. Wikipedia. Quantum circuit (2018). https://en.wikipedia.org/wiki/Quantum_circuit
40. M.A. Nielsen, I.L. Chuang, *Quantum Computation and Quantum Information*, ed. by M.A. Nielsen, I.L. Chuang (Cambridge University Press, Cambridge, 2010)
41. A. Barenco, C.H. Bennett, R. Cleve, D.P. DiVincenzo, N. Margolus, P. Shor, T. Sleator, J.A. Smolin, H. Weinfurter, Phys. Rev. A **52**(5), 3457 (1995)
42. C. Rigetti, M. Devoret, Phys. Rev. B **81**(13), 134507 (2010)
43. S. Sheldon, E. Magesan, J.M. Chow, J.M. Gambetta, Phys. Rev. A **93**(6), 060302 (2016)
44. IBM. IBM Q Experience Device (2018). https://www.research.ibm.com/ibm-q/technology/devices/
45. J. Kelly. A Preview of Bristlecone, Google's New Quantum Processor (2018). https://ai.googleblog.com/2018/03/a-preview-of-bristlecone-googles-new.html
46. D. Maslov, S.M. Falconer, M. Mosca, IEEE Trans. Comput. Aided Design Integr. Circuits Syst. **27**(4), 752 (2008)
47. J.B. Hertzberg, E.J. Zhang, S. Rosenblatt, E. Magesan, J.A. Smolin, J.B. Yau, V.P. Adiga, M. Sandberg, M. Brink, J.M. Chow, et al., Preprint, arXiv:2009.00781 (2020)
48. R. Wille, D. Große, L. Teuber, G.W. Dueck, R. Drechsler, in *38th International Symposium on Multiple Valued Logic, 2008. ISMVL 2008.* (IEEE, Piscataway, 2008), pp. 220–225
49. H. Abraham, A. Offei, R. Agarwal, I.Y. Akhalwaya, G. Aleksandrowicz, T. Alexander, M. Amy, E. Arbel, Arijit, A. Asfaw, A. Avkhadiev, C. Azaustre, AzizNgoueya, A. Banerjee, A. Bansal, P. Barkoutsos, A. Barnawal, G. Barron, G.S. Barron, L. Bello, Y. Ben-Haim, D. Bevenius, A. Bhobe, L.S. Bishop, C. Blank, S. Bolos, S. Bosch, Brandon, S. Bravyi, Bryce-Fuller, D. Bucher, A. Burov, F. Cabrera, P. Calpin, L. Capelluto, J. Carballo, G. Carrascal, A. Chen, C.F. Chen, E. Chen, J.C. Chen, R. Chen, J.M. Chow, S. Churchill, C. Claus, C. Clauss, R. Cocking, F. Correa, A.J. Cross, A.W. Cross, S. Cross, J. Cruz-Benito, C. Culver, A.D. Córcoles-Gonzales, S. Dague, T.E. Dandachi, M. Daniels, M. Dartiailh, D. Frr, A.R. Davila, A. Dekusar, D. Ding, J. Doi, E. Drechsler, Drew, E. Dumitrescu, K. Dumon, I. Duran, K. EL-Safty, E. Eastman, G. Eberle, P. Eendebak, D. Egger, M. Everitt, P.M. Fernández, A.H. Ferrera, R. Fouilland, F. Chevallier, A. Frisch, A. Fuhrer, B. Fuller, M. George, J. Gacon, B.G. Gago, C. Gambella, J.M. Gambetta, A. Gammanpila, L. Garcia, T. Garg, S. Garion, A. Gilliam, A. Giridharan, J. Gomez-Mosquera, Gonzalo, S. de la Puente González, J. Gorzinski, I. Gould, D. Greenberg, D. Grinko, W. Guan, J.A. Gunnels, M. Haglund, I. Haide, I. Hamamura, O.C. Hamido, F. Harkins, V. Havlicek, J. Hellmers, Ł. Herok, S. Hillmich, H. Horii, C. Howington, S. Hu, W. Hu, J. Huang, R. Huisman, H. Imai, T. Imamichi, K. Ishizaki, R. Iten, T. Itoko, J. Seaward, A. Javadi, A. Javadi-Abhari, W. Javed, Jessica, M. Jivrajani, K. Johns, S. Johnstun, Jonathan-Shoemaker, V. K, T. Kachmann, A. Kale, N. Kanazawa, Kang-Bae, A. Karazeev, P. Kassebaum, J. Kelso, S. King, Knabberjoe, Y. Kobayashi, A. Kovyrshin, R. Krishnakumar, V. Krishnan, K. Krsulich, P. Kumkar, G. Kus, R. LaRose, E. Lacal, R. Lambert, J. Lapeyre, J. Latone, S. Lawrence, C. Lee, G. Li, D. Liu, P. Liu, Y. Maeng, K. Majmudar, A. Malyshev, J. Manela, J. Marecek, M. Marques, D. Maslov, D. Mathews, A. Matsuo, D.T. McClure, C. McGarry, D. McKay, D. McPherson, S. Meesala, T. Metcalfe, M. Mevissen, A. Meyer, A. Mezzacapo, R. Midha, Z. Minev, A. Mitchell, N. Moll, J. Montanez, G. Monteiro, M.D. Mooring, R. Morales, N. Moran, M. Motta, M. F, P. Murali, J. Müggenburg, D. Nadlinger, K. Nakanishi, G. Nannicini, P. Nation, E. Navarro, Y. Naveh, S.W. Neagle, P. Neuweiler, J. Nicander, P. Niroula, H. Norlen, NuoWenLei, L.J. O'Riordan, O. Ogunbayo, P. Ollitrault, R. Otaolea, S. Oud, D. Padilha, H. Paik, S. Pal, Y. Pang, V.R. Pascuzzi, S. Perriello, A. Phan, F. Piro, M. Pistoia, C. Piveteau, P. Pocreau, A. Pozas-

Kerstjens, M. Prokop, V. Prutyanov, D. Puzzuoli, J. Pérez, Quintiii, R.I. Rahman, A. Raja, N. Ramagiri, A. Rao, R. Raymond, R.M.C. Redondo, M. Reuter, J. Rice, M. Riedemann, M.L. Rocca, D.M. Rodríguez, RohithKarur, M. Rossmannek, M. Ryu, T. Sapv, S. Ferracin, M. Sandberg, H. Sandesara, R. Sapra, H. Sargsyan, A. Sarkar, N. Sathaye, B. Schmitt, C. Schnabel, Z. Schoenfeld, T.L. Scholten, E. Schoute, J. Schwarm, I.F. Sertage, K. Setia, N. Shammah, Y. Shi, A. Silva, A. Simonetto, N. Singstock, Y. Siraichi, I. Sitdikov, S. Sivarajah, M.B. Sletfjerding, J.A. Smolin, M. Soeken, I.O. Sokolov, I. Sokolov, S. Thomas, Starfish, D. Steenken, M. Stypulkoski, S. Sun, K.J. Sung, H. Takahashi, T. Takawale, I. Tavernelli, C. Taylor, P. Taylour, S. Thomas, M. Tillet, M. Tod, M. Tomasik, E. de la Torre, K. Trabing, M. Treinish, T. Pe, D. Tulsi, W. Turner, Y. Vaknin, C.R. Valcarce, F. Varchon, A.C. Vazquez, V. Villar, D. Vogt-Lee, C. Vuillot, J. Weaver, J. Weidenfeller, R. Wieczorek, J.A. Wildstrom, E. Winston, J.J. Woehr, S. Woerner, R. Woo, C.J. Wood, R. Wood, S. Wood, S. Wood, J. Wootton, D. Yeralin, D. Yonge-Mallo, R. Young, J. Yu, C. Zachow, L. Zdanski, H. Zhang, C. Zoufal, M. Čepulkovskis. Qiskit: an open-source framework for quantum computing (2019). https://doi.org/10.5281/zenodo.2562110

50. M. Hutchings, J.B. Hertzberg, Y. Liu, N.T. Bronn, G.A. Keefe, M. Brink, J.M. Chow, B. Plourde, Phys. Rev. Appl. **8**(4), 044003 (2017)

51. S. Rosenblatt, J. Hertzberg, M. Brink, J. Chow, J. Gambetta, Z. Leng, A. Houck, J. Nelson, B. Plourde, X. Wu, et al., in *APS Meeting Abstracts* (2017)

52. J. Ghosh, A. Galiautdinov, Z. Zhou, A.N. Korotkov, J.M. Martinis, M.R. Geller, Phys. Rev. A **87**(2), 022309 (2013)

53. Rigetti. The Quantum Processing Unit (QPU) (2018). https://www.rigetti.com/qpu

54. D-Wave Systems Inc. D-Wave System Documentation (2018). https://docs.dwavesys.com/docs/latest/

55. E. Farhi, J. Goldstone, S. Gutmann, M. Sipser, Preprint. quant-ph/0001106 (2000)

56. D. Venturelli, M. Do, E. Rieffel, J. Frank, Quant. Sci. Technol. **3**(2), 025004 (2018)

57. A. Shafaei, M. Saeedi, M. Pedram, in *Design Automation Conference (ASP-DAC), 2014 19th Asia and South Pacific* (IEEE, Piscataway, 2014), pp. 495–500

58. S.S. Tannu, M.K. Qureshi, in *Proceedings of the Twenty-Fourth International Conference on Architectural Support for Programming Languages and Operating Systems* (ACM, New York, 2019), pp. 987–999

59. A. Ash-Saki, M. Alam, S. Ghosh, in *Proceedings of the 56th Annual Design Automation Conference 2019* (ACM, New York, 2019), p. 141

Quantum True Random Number Generator

Abdullah Ash Saki, Mahabubul Alam, and Swaroop Ghosh

1 Introduction

Quantum computers are well-suited for true random number generation due the presence of *superposition* property. A quantum bit or qubit can be constructed with equal probabilities in the superposition of $|0\rangle$ and $|1\rangle$, i.e., $|\psi\rangle = (|0\rangle + |1\rangle)/\sqrt{2}$. As a result, reading this qubit multiple times will produce 0s and 1s with equal probability. This insight can be used to create a TRNG. It may seem like a simple concept, but contemporary quantum computers have noise concerns that complicate matters.

1.1 Motivating Study

Figure 1a illustrates a TRNG quantum circuit. It entails using a single-qubit $RY(\pi/2)$ gate on a qubit in the 0 state. The RY gate rotates the Bloch sphere by 90° degrees around the Y-axis, putting the qubit in equal superposition. Note that we can utilize each qubit in a quantum computer, e.g., IBMQX2 (QX2), IBMQX4 (QX4), and IBMQ_16_Melbourne (IBMQ16), (Fig. 1b) to generate random bits. In a superposition state, P(0) and P(1) are equal (Fig. 1c). However, present quantum computers are noisy and prone to several errors, including gate error and readout error (noise and error are used interchangeably in this chapter). Figure 1d and e depicts the single-qubit gate error and readout error for each qubit in IBMQX4 collected over 43 days. The plots signify that the qubits are erroneous, and the

A. A. Saki (✉) · M. Alam · S. Ghosh
Pennsylvania State University, University Park, PA, USA
e-mail: ash.saki@live.com; mxa890@psu.edu; szg212@psu.edu

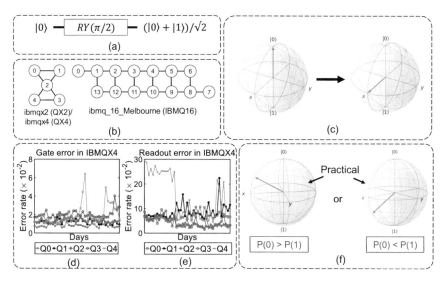

Fig. 1 (**a**) Circuit for creating superposition state, (**b**) qubit connectivity graph of IBMQX2, IBMQX4, and IBMQ16, (**c**) Bloch sphere representation of the superposition state of the qubit in ideal case such that P(0) = P(1). Temporal and spatial variation of (**d**) gate error and (**e**) readout/measurement error, (**f**) Skewed qubit state due to various errors

error rates demonstrate both spatial and temporal variation. These errors cause the superposition state to depart from the ideal situation, resulting in an uneven probability of $|0\rangle$ and $|1\rangle$, as conceptually displayed by Bloch sphere Fig. 1f. Experimental results also corroborate this intuition. The ideal TRNG circuit in Fig. 1a, when ran on Q1 of a real quantum computer IBMQX4, generated highly uneven 1/0 ratio (Fig. 2a). Additionally, we undertake quantum state tomography (QST) [1] experiments on IBMQX4's qubit Q1 and reconstruct the qubit state's density matrix. In Fig. 2b, the real-part of the reconstructed density matrix is displayed. The diagonal elements of the density matrix represent the probability of the pure states $|0\rangle$ and $|1\rangle$ and should ideally be 0.5 for the equal superposition state (Due to a finite number of samples such as 8192 shots used in the experiments, it may not be exactly 0.5). The diagonal elements, on the other hand, depart significantly from their expected values of 0.5 each, with the likelihood of 0 being significantly larger than 1. It appears that this observation is consistent with the highly skewed (less than 1) 1-to-0 ratio depicted in Fig. 2a. Additionally, we execute QST on two more publicly accessible IBM quantum computers (Q0 of IBMQX2 and Q2 of IBMQ16) and reconstruct the density matrix in each case (Fig. 2b). As expected, the diagonal elements diverge from the expected value of 0.5 (the amount of deviation varies due to error rate variations among devices).

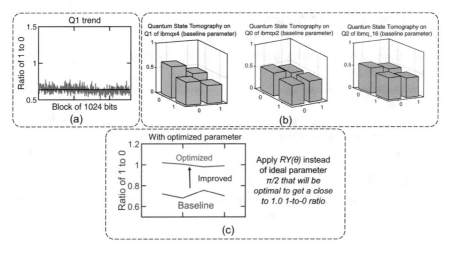

Fig. 2 (**a**) *Experimental* 1-to-0 ratio from Qubit-1 (Q1) of IBMQX4 showing major deviation from the ideal value of 1.0, (**b**) reconstructed density matrix using quantum state tomography performed on Q1 of IBMQX4, Q0 of IBMQX2, and Q2 of IBMQ16, and (**c**) conceptual depiction of the proposed idea where a gate parameter other than the ideal value of $\pi/2$ is used to restore the 1-to-0 ratio close to 1.0

1.2 Proposed Approaches

The above discussion indicates that obtaining dependable random numbers from near-term QC is a difficult problem. With the aid of the Bloch sphere, we can conceptualize the problem and design a solution. The imbalance in the 1/0 ratio tells us that a $\pi/2$ rotation along Y-axis is not placing the qubit at the equator of the Bloch sphere, which is the equal superposition state. Intuitively, a different rotation angle than $\pi/2$ may compensate for the error and place the qubit in an equal superposition (Fig. 2c). On the other hand, the search for the optimal parameter value necessitates systematic exploration procedures. One such approach, used in this chapter, is the *hardware-in-the-loop* technique where a TRNG circuit on a QC and an optimizer on a classical computer iterates in a loop. The loop continues until a gate parameter results in a 1/0 ratio close to 1. While the hardware-in-the-loop technique provides optimal parameter selection, the loop may require several iterations to converge. We incorporate a regression-based machine learning algorithm to expedite the search process. The model accepts error rates as input and outputs the optimal gate parameter value. The model is developed using synthetic training data obtained through error characterization experiments and simulations, which are designed to quantify the TRNG's sensitivity to various noise sources. Additionally, we use a combination of hardware-in-the-loop and machine learning to further optimize the parameter.

1.3 Chapter Contribution

- We evaluate several noise processes to determine their actual impacts on the rotation-based TRNG.
- We propose optimizing quantum gate parameters to adjust for nefarious noise effects.
- We develop a regression-based machine learning model to predict optimized gate parameters and minimize classical optimization overhead.
- We propose the fusion of machine learning-based prediction with hybrid quantum-classical optimization to fine-tune the gate parameter for best results, and finally,
- We run tests from NIST statistical test suite to quantify the quality improvement of TRNs with the proposed approach.

1.4 Chapter Organization

The rest of the chapter is organized as follows: Sect. 2 presents the characterization of various noise sources for TRNG application using real-device experiments and simulations. The details of machine learning assisted parameter optimization methodology along with the quantum-classical hybrid optimization approach are elucidated in Sect. 3. The key results and discussions are presented in Sect. 4. Conclusions are drawn in Sect. 5 along with future outlook.

2 Characterizing Noise for TRNG Application

2.1 Readout Error

Accurately characterizing the readout/measurement error requires conducting a series of experiments and creating a measurement error characterization matrix (R) from the findings. As readout errors are *bit-flip* in nature, the characterization involves finding the bit-flip probabilities of states $|1\rangle$ and $|0\rangle$, for a single qubit. Thus, the measurement error characterization matrix for a single qubit is a 2×2 matrix as shown in Fig. 3a. An element of the matrix, M_{xy}, means the probability of measuring state "x" while the qubit is prepared in state "y."

To find the bit-flip probabilities for a single qubit, the qubit is repeatedly prepared in states $|1\rangle$ and $|0\rangle$ and measured. Circuit—I and II in Fig. 3a show the circuits for preparing a qubit in states $|1\rangle$ and $|0\rangle$, respectively. In IBM systems, a qubit always starts from the $|0\rangle$ state. Therefore, to prepare the qubit in state $|1\rangle$, a quantum NOT (X) gate is applied on the qubit in the circuit—I in Fig. 3a. Suppose a qubit is prepared state $|1\rangle$ and measured 8192 times out of which 1 is recorded 7000 times

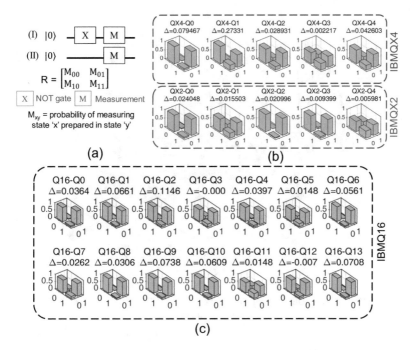

Fig. 3 (**a**) Circuits for characterizing readout error and experimental readout characterization matrices for each qubit of (**b**) QX4, QX2, and (**c**) Q16. The results show qubit-to-qubit and chip-to-chip variations. For example, 1s flip more than 0s in Q1 of QX4, which supports a ratio biased to 0 in Fig. 1g. (Δ = Diff. of diagonals)

and 0 is recorded 1192 times. Therefore, the bit-flip probability of state $|1\rangle$, i.e., M_{01}, will be $1192/8192 \approx 0.15$ (similarly, M_{11} for this example is $7000/8192$ or $1 - M_{01} \approx 0.85$).

Figure 3b and 3c show the measurement characterization matrix (R) for each qubit of 3 IBM machines, namely, IBMQX4, IBMQX2, and IBMQ16.

Note that the data from the circuit—I is affected by both gate error (from the X gate) and readout error. As the circuit comprises just 1 single-qubit gate, and single-qubit gates in IBM systems have a substantially lower error rate, we may infer that the divergence in data is largely due to readout error.

2.1.1 Skew in 1-to-0 Ratio Due to Readout Error

For the TRNG application, the difference between M_{00} and M_{11} (not the absolute values) dictates the asymmetry in the $1/0$ ratio. When the difference is small, both 1 and 0 flip with equal (or close to equal) probabilities leading to an (close to) equal number of 1s and 0s.

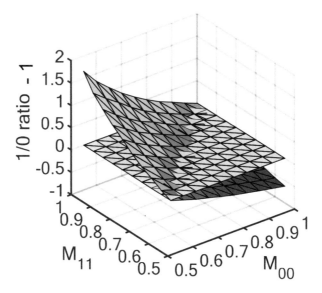

Fig. 4 Simulated results with varying M_{00} and M_{11}. The 1/0 ratio skews more if the difference between M_{00} and M_{11} is large. The surface at z = 0 is for reference

We sweep the values of M_{00} and M_{11} over a 2D grid from 0.55 to 1.0 each and compute the 1/0 ratio (Fig. 4) using IBM's QASM simulator [2]. The result shows that the ratio is 1.0 when $M_{00} = M_{11}$ and the ratio deviates from 1.0 as the difference between M_{00} and M_{11} increases.

2.2 Gate Error and Decoherence

In a quantum computer, the TRNG circuit is composed of a single-qubit gate, for example, $RY(\theta)$ on a qubit. The rotation gate $RY(\theta)$ is implemented using microwave pulses in IBM systems with superconducting qubits. The rotation angle is determined by the amplitude of the pulse (also by pulse duration and shape). Due to control mechanism faults, the amplitude is not always accurate, resulting in a rotation error (i.e., the real rotation becomes $\theta \pm \delta$). Table 1 summarizes experimental rotation errors obtained using Qiskit [2] on IBM systems (QX4, QX2, and IBMQ16) (averaged over 123 experiments on each computer).

Additionally, the qubit experiences decoherence (T1-relaxation and T2-dephasing). Theoretically, the T1 relaxation can dampen $|1\rangle$ state and distort the 1/0 ratio. However, the damping likelihood is extremely low for a single gate period, even for the worst known T1 time of 23μs measured on 24-July-2019 $((p \approx 1 - exp(-t_{gate}/T_1) = 1 - exp(-50ns/23\mu s) \approx 0.002)$. Notably, T2 dephasing does not affect TRNG since it simply shifts one equal superposition state onto another equal superposition state on the Bloch sphere.

Table 1 Experimental rotation-error/gate of $RY(\pi/2)$ gate for QX4, QX2, and IBMQ16

Qubit (Q) #	QX4 ($\times 10^{-3}$ rad)	QX2 (rad)	Qubit (Q) #	Q16 (rad)	Qubit (Q) #	Q16 (rad)	Qubit (Q) #	Q16 (rad)
0	0.2472	−0.0013	0	−0.0009	5	0.0002	10	0.0005
1	−0.2378	−0.0006	1	−0.0011	6	−0.0001	11	−0.0006
2	−0.1050	−0.0023	2	−0.0008	7	−0.0002	12	−0.0012
3	−0.2008	−0.0016	3	−0.0003	8	−0.0006	13	−0.0002
4	−0.1088	−0.0009	4	−0.0011	9	0.0005		

Fig. 5 Simulated 1/0 ratio with rotation error ($T1 = 23\mu s$)

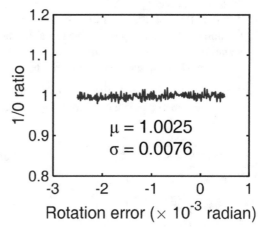

2.2.1 Minor Deviation in 1/0 Ratio Due to Gate Error and Decoherence

The minor impact of gate error and decoherence on the TRNG application can be validated through simulation. We use the IBM QASM Simulator to model the effect of rotation errors with the worst T1-relaxation on the 1/0 ratio, and the results are plotted in Fig. 5. Due to the low error values associated with a single-qubit gate operation, the 1/0 ratio does not stray significantly from the ideal value of 1.

Take note that the TRNG circuit has only one single-qubit gate, and that readout error is the most prevalent error mode in this TRNG application.

3 Parameter Optimization to Mitigate the Noise Effect

3.1 Reducing Bias in TRNG—Traditional Approach

Because of the readout error outlined in Sect. 2, a quantum TRNG can provide a strongly biased output (skewed 1/0 ratio). Post-processing using standard bias correction algorithms (e.g., Von Neumann correction) is commonly employed to establish a balanced distribution of 1s and 0s from biased hardware TRNGs [3–

5]. A Von Neumann corrector translates the bit pair [0, 1] to an output 1, [1, 0] to an output 0, and neither [0, 0] nor [1, 1] to an output. Such post-processing is not appropriate for a quantum TRNG due to the following reasons:

- **Throughput reduction**: Classical post-processing decreases bias at the expense of performance by converting bit pairs to single bits and discarding many bit pairs. Von Neumann correction was done to 81,920 bits derived from IBMQX4's Q1 and Q4 with a 1/0 ratio of 0.57. Following post-processing, we obtained a 20,464-bit bitstream with a 1/0 ratio of 0.991. However, it resulted in a reduction of the output bits by $\approx 75\%$.

- **Temporal and spatial variation in throughput:** As depicted in Fig. 1d–e, errors in qubits show both spatial (qubit-to-qubit) and temporal variation. Due to these variations, the post-processing step also suffers which may lead to variables throughput at different times for different qubits.

The suggested solutions (Fig. 6), which will be described next, aim to overcome the aforementioned drawbacks of traditional post-processing.

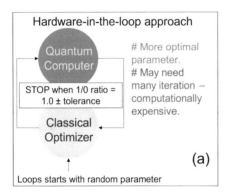

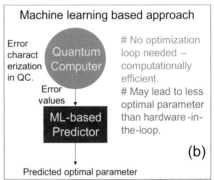

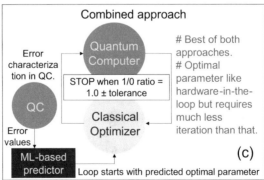

Fig. 6 Overview and simplified flow of proposed methods for finding the optimal gate parameter. The parameter from these methods can reduce noise-induced bias in TRNG without affecting the throughput. (**a**) Hardware-in-the-loop approach, (**b**) machine learning-based approach, and (**c**) combined approach

3.2 Hardware-in-the-Loop Approach

In the hardware-in-the-loop approach, we employ a closed-loop gate parameter optimization protocol where a quantum computer and classical optimizer iterate in a loop to fix the gate parameter to remove the bias in the 1/0 ratio. Figure 6a visualizes the idea. In this approach, the TRNG circuit in a real quantum computer starts with an initial gate parameter (i.e., the rotation angle of the gate $RY(\theta)$) and generates a bitstream. Next, a classical optimizer calculates the 1/0 ratio from the bitstream and generates a new gate parameter to minimize the objective function $f = (1 - ratio)^2$ (i.e., tries to make the ratio close to 1.0). The routine continues until the 1/0 ratio reaches a set tolerance limit. For best results, we use a global optimizer (differential-evolution [6]) available from the SciPy Python package. The disadvantage of the preceding approach is that the parameter adjusted on one day may not perform best on another owing to temporal variation (Fig. 1d–e). Additionally, each quantum computer may have unique readout noise characteristics. As a result, hardware-in-the-loop must be invoked for each device each time. It will be computationally expensive, as the classical optimizer will have to search a large parameter space each time, and many experiments on the real QC will have to be scheduled. As a result, a more generic and efficient strategy is required that can account for temporal, spatial, and device-to-device changes in errors while minimizing loop iterations.

3.3 Machine Learning-Based Approach

The machine learning-based approach is illustrated in Fig. 6b, where a trained predictor model outputs the optimal gate parameter based on the device noise values instead of running hybrid loop with real device. To understand the construction of the predictor model, we return to the relation between error characteristics and the skewed 1/0 ratio discussed in Sect. 2. Due to limited public access, gathering enough data from IBM's real devices to develop a generic statistical model is time-consuming. As a result, we build the model using simulation. To generate the training data, we run the hybrid quantum-classical optimization loop, but this time with the noisy QASM simulator rather than the real QC. We load the gate error, T1 relaxation, and readout error models into the simulator and sweep the values of gate (rotation) error from -0.01 to 0.01, T1 time from $20\mu s$ to $70\mu s$, and M_{00} and M_{11} from 0.55 to 1.0 each. These values encompass the entire range of real-world device specifications. The classical optimizer is set to search in a range of 0 to 2π with a tolerance of 10^{-4} for the objective function. Figure 7, for example, depicts the trend of the optimized gate parameter value as M_{00} and M_{11} (readout error) are varied. To compensate for the error, the parameter deviates more from the ideal value of $\pi/2$ as the difference between M_{00} and M_{11} grows larger. It should be noted that a higher-dimensional trend that includes all errors cannot be visually represented.

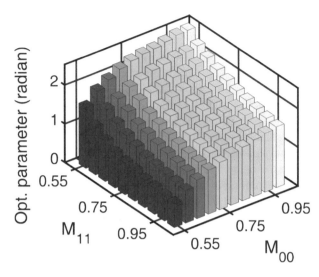

Fig. 7 Optimized parameter value with respect to readout errors

The acquired data is then used to train a K-nearest neighbor regression model. The features used in training are the gate rotation error, the T1 time, and the readout error (M_{00} and M_{11}), whereas the output variable is the optimized gate parameter. The gate parameter is specified in radians. We employ a neighbor size of three to forecast the ideal parameter value for a qubit based on its error characteristics. We compare the regression-based optimized parameter values to the raw hardware-in-the-loop parameter values using IBMQX4, IBMQX2, and IBMQ16 as a sanity check (Table 2). The comparison demonstrates that the average deviation from the hardware-in-the-loop approach is less than 3%, with QX4 and QX2 having errors of less than 1% as tabulated in Table 2.

3.4 Combined Approach: Machine Learning and Hardware-in-the-Loop

The hardware-in-the-loop and the machine learning approaches can be combined to get a fine-tuned gate parameter (Fig. 6c). First, the regression-based model predicts a parameter value. Next, the hardware-in-the-loop starts from this value, and the classical optimizer searches in a narrow range in the vicinity of the predicted parameter to further fine-tune it. In our experiments, we confine the range between ±0.1 of the predicted parameter value. In this way, we can find a fine-tuned gate parameter like the hardware-in-the-loop approach, but much faster than before.

Table 2 Comparison between the parameter values predicted by the model and values computed using a real device in the loop

Q	QX4			QX2			Q16 (Q0-Q4)		
	Pred.	HW	% err	Pred.	HW	% err	Pred.	HW	% err
0	1.6540	1.6893	1.7810	1.6071	1.5945	0.7848	1.6177	1.5684	2.8685
1	1.9765	1.9797	0.1635	1.6076	1.6118	0.2598	1.6407	1.5910	2.8841
2	1.6158	1.6237	0.4002	1.5947	1.6031	0.5196	1.6379	1.7213	4.8466
3	1.6317	1.5999	1.6066	1.5665	1.5788	0.7681	1.6515	1.572	5.0572
4	1.6558	1.6741	0.9222	1.5659	1.5406	1.5695	1.6162	1.6296	0.7775
		Avg.	0.9747		Avg.	0.7804			

Q	Q16 (Q5-Q7)			Q	Q16 (Q8-10)			Q	Q16 (Q11-Q13)		
	Pred.	HW	% err		Pred.	HW	% err		Pred.	HW	% err
5	1.6181	1.6389	1.2056	8	1.5986	1.6519	3.0993	11	1.6292	1.6684	2.2817
6	1.6180	1.6126	0.3123	9	1.6040	1.6533	2.8675	12	1.5837	1.5961	0.7178
7	1.6552	1.5703	4.9299	10	1.6379	1.6213	0.9682	13	1.6407	1.6092	1.8278
										Avg.	2.6507

4 Results and Discussions

4.1 Optimal Parameter Search Overhead

Before generating random numbers, the quantum TRNG runs multiple times to determine the ideal parameter value. We name the number of calls as the optimal parameter search overhead. Using the hardware-in-the-loop technique, we find that the quantum computer (QC) is invoked on average $\approx$ 23 times for a single qubit. For approaches based on machine learning, the QC is invoked a fixed three times (once each for readout error, gate error, and T1 time). Finally, the combined technique has an average overhead of $\approx$ 7, which is the total of QC calls for machine learning-based prediction (i.e., 3) + QC calls for the optimization loop ($\approx$ 4 on average). As this technique begins with a near-optimal parameter value, it results in significantly faster loop convergence. It should be noted that the ML model has a one-time training cost. The training is done with simulated data. It only takes a few minutes to generate this training data and then train the model. For example, on a Linux (Ubuntu 18.04) virtual machine on Win 10 host with a core i7 − 6700 3.40GHz (2 cores) processor and 4 GB of RAM, training data generation took about 258 seconds and model training and parameter prediction took about 3 seconds with an error-parameter sweeping granularity of 0.05. Because it is a one-time cost, the training data and/or model can be reused for all future instances without modification, making the ML-based approach faster than the raw hardware-in-the-loop approach.

4.2   Improvement of 1/0 Ratio

The proposed approaches can significantly improve 1/0 ratio compared to baseline quantum TRNG circuit, and experimental results from three IBM quantum computers corroborate to that claim. The experimental 1/0 ratio from three IBM devices are plotted in Fig. 8a–c. The results contain 1/0 ratio for three cases, namely *baseline* (i.e., TRNG circuit with a $RY(\pi/2)$ gate), *optimized* (i.e., TRNG circuit with a $RY(\theta)$ gate, where the parameter θ is computed from the machine learning-based predictor model), and *optimized-fine or opt.-fine* (i.e., TRNG circuit with a $RY(\theta)$ gate, where the parameter θ is computed from the combined approach). As evident from experimental data in Fig. 8, both approaches *optimized* and *optimized-fine* perform better than the *baseline* TRNG providing better 1/0 ratio in terms of mean and standard deviation. From Fig. 8d, the average improvements of the mean and standard deviation are 22.14%, 3.78% and 10.05%, and 97.86%, 91.7%, and 92%, respectively. Most importantly, the proposed approaches can improve the worst performing qubits (88.57%, 9.60% and 26.83% for each chip (baseline vs. opt.-fine). Note that, 1/0 ratios are computed from 204, 800 bits/qubit.

To confirm the improvement further, we execute QST on each qubit of the computers and plot the results in Figs. 9 and 10. The results demonstrate a very minor difference in the diagonal elements of the density matrices, which represent the (almost) equal probability of 1s and 0s required for randomization. For instance, the difference between diagonal elements (i.e., probabilities of 0 and 1) in Q2 of IBMQ16 with the baseline parameter is 0.1220 (Fig. 1), whereas the new difference is 0.0073 (Fig. 10), demonstrating the effectiveness. (Note that we applied our approaches to generate bits from each qubit separately; not from all qubits at once.)

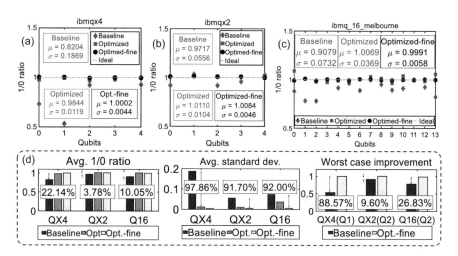

Fig. 8 1/0 ratio from, (**a**) QX4, (**b**) QX2, and, (**c**) IBMQ16. (**d**) The average mean and standard deviation and worst-case mean improvements of 1/0 ratio

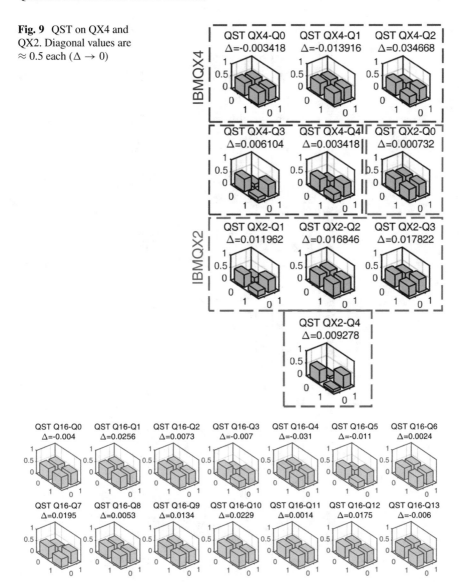

Fig. 9 QST on QX4 and QX2. Diagonal values are ≈ 0.5 each (Δ → 0)

Fig. 10 QST on each qubit of IBMQ16

4.3 NIST Test Suite Results

We run several tests from the NIST statistical test suite [7] multiple times at a significance level of 0.01 to validate the randomness of the generated bits. Table 3 summarizes the results of the tests (cells with value greater than 0.01 mean PASS and value less than 0.01 mean FAIL). While the bitstream generated by the baseline

Table 3 NIST results. Each run is conducted with data collected on different days from IBMQX4. QX4 is chosen as it has the worst qubits

NIST Test Names	Baseline p-value				Proposed method p-value			
	Run-1	Run-2	Run-3	Run-4	Run-1	Run-2	Run-3	Run-4
Frequency Test (Mono-bit)	2.1e−70	2e−113	1e−100	3e−119	0.0287	0.8512	0.3648	0.4749
Frequency Test within a Block	6.7e−26	1.3e−42	1.9e−47	3.7e−48	0.2264	0.3936	0.5083	0.8051
Run Test	0	0	0	0	0.0293	0.9623	0.6798	0.6900
Longest Run of Ones in a Block	4.3e−9	8.3e−50	3e−10	1.6e−17	0.2261	0.9700	0.3710	0.4713
Binary Matrix Rank Test	0.6221	0.1974	0.6932	0.4585	0.3335	0.6132	0.5993	0.2311
Discrete Fourier Transform (Spectral)	0.4682	4.4e−32	0.0004	0.0032	0.3018	0.2396	0.4472	0.6992
Approximate Entropy Test	3.7e−9	1.3e−95	5.2e−27	1.7e−28	0.7963	0.0103	0.1490	0.5809
Cumulative Sums (Forward) Test	9.7e−71	5e−114	6e−101	8e−120	0.0227	0.6149	0.4165	0.4504
Cumulative Sums (Reverse) Test	8.9e−71	1e−114	6e−101	1e−119	0.0574	0.4572	0.7165	0.4456

Table 4 Comparison among emerging technology-based TRNGs

Technology	Entropy source	Bit-rate	Entropy (p-value)
CMOS [8]	Oscillator jitter	Mbps	0.3240
Memristor [9]	Stochastic diffusion	Kbps	0.0098
STT-MRAM [10]	Stochastic STT switching	10^{-1} Mbps	> 0.0001
Quantum optics [11]	Raman scattering	Kbps–Mbps	0.7197
Superconducting qubit (This work)	Quantum superposition	Mbps (w/active reset [12])	> 0.7963 (Run-1)

implementation fails in several tests, the bitstream generated by our method passes all of them, which demonstrates the improvement in randomness. The ML-based approach is used to generate the bitstream for Runs 1 and 3, and the combined approach is used to generate the bitstream for Runs 2 and 4. It should be noted that the raw hardware-in-the-loop approach is not used in this case because the combined approach incorporates it.

4.4 Comparative Analysis

Table 4 compares the proposed TRNG against a number of emerging technology-based TRNGs, including CMOS [8], memristor [9], STT-MRAM [10], and quantum random number generators [11]. The suggested TRNG can attain a bit rate of up to Mbps per qubit and the throughput scales linearly with the number of qubits. However, the hardware in the preceding studies is application-specific, whereas the suggested implementation is based on a universal quantum computer that includes true random number generation as one of its applications.

4.5 Related Works

Several other works from academia and industry focus on generating true random number from gate-based quantum computers. AWS Braket demonstrated random number generation by using two quantum computers [13]. In their approach, they generated raw bitstrings from two different QCs. Next, they fed the raw strings in a classical randomness extractor. The details of the approach can be found in Ref. [14].

Cambridge Quantum launched a service named IronBridge that generates cryptographically safe random numbers using quantum computers [15]. In their approach, they use a three-qubit Bell test to evaluate the Mermin correlator (M) [16], which outputs a value between zero and four. A value above two indicates quantum entanglement. The value works as a self-test as it quantifies the performance of the device and the amount of randomness. The protocol is aborted if the value is less than two. A tutorial on IronBridge circuits can be found in Ref. [17].

Another software tool named QRAND [18] offers multi-protocol and multiple platforms support for quantum true random number generation, It offers simple Hadamard gate-based and more complex quantum entanglement-based [19] protocols to generate true random numbers.

5   Conclusion and Future Outlook

We present a TRNG using superconducting qubit-based universal quantum computers. We show that the theoretical TRNG circuit does not work due to various errors present in the near-term quantum computers by performing experiments and simulations to characterize the effects of the noise on TRNG. We employ a quantum-classical gate parameter optimization protocol to correct the error-induced bias. We propose a machine learning-based approach to accelerate the parameter optimization process. Finally, we demonstrate the validity of our proposal with the NIST test suite.

5.1   Future Outlook

As a future research direction, crosstalk-compensated and drift-free random number generation need to be explored. The proposed approach uses a single qubit from a quantum computer to generate random numbers. However, if TRNG circuits are run on every qubit on hardware simultaneously, they may suffer from crosstalk from neighboring qubits, especially the always-on ZZ-crosstalk in Transmon-type superconducting qubits [20]. Therefore, the effect of crosstalk on true random generation can be studied, and crosstalk can be included, in addition to readout error, in the statistical model for optimal parameter generation. The ability to account for crosstalk and use all qubits of the device simultaneously can increase the throughput of random number generation multiple folds.

Besides crosstalk, managing *drift* for random number generation can be studied. The error rates such as readout error, gate error, and T1-relaxation time change over time which is known as drift. Due to drift, the optimal gate parameter needs to be updated with time. The update interval can be explored and efficient ways to tackle this drift need investigation.

A detailed report on practical considerations and use cases of quantum true random number generated can be found in [21]. The report discusses the features and advantages of quantum true random number generators and considers scenarios, where using such random number generators may be worthwhile.

Finally, random numbers are critical for many applications in fields like computer science, finance, cryptography, and cybersecurity. Quantum computers are ideal for this purpose. Exploring other protocols, in addition to using Hadamard gate and quantum entanglement, and techniques to mitigate noise are necessary to generate more reliable and cryptographically secure true random numbers.

Acknowledgments This work is supported by the NSF (DGE-2113839, OIA-2040667, and CNS-2129675) and seed grants from the Penn State ICDS and the Huck Institute of Life Sciences.

We acknowledge the use of IBM Quantum services for this work. The views expressed are those of the authors and do not reflect the official policy or position of IBM or the IBM Quantum team.

References

1. J.A. Smolin, J.M. Gambetta, G. Smith, Efficient method for computing the maximum-likelihood quantum state from measurements with additive gaussian noise. Phys. Rev. Lett. **108**(7), 070502 (2012)
2. M. Treinish et al., Qiskit: An open-source framework for quantum computing (2019)
3. J. Von Neumann, 13. various techniques used in connection with random digits. Appl. Math. Ser. **12**, 5 (1951)
4. B. Jun, P. Kocher, The intel random number generator. Cryptography Research Inc., White paper, vol. 27, pp. 1–8 (1999)
5. V. Rožić, B. Yang, W. Dehaene, I. Verbauwhede, Iterating von Neumann's post-processing under hardware constraints, in *2016 IEEE International Symposium on Hardware Oriented Security and Trust (HOST)* (IEEE, 2016), pp. 37–42
6. R. Storn, K. Price, Differential evolution—a simple and efficient heuristic for global optimization over continuous spaces. J. Global Optim. **11**(4), 341–359 (1997)
7. L.E. Bassham, III et al., SP 800-22 rev. 1a. a statistical test suite for random and pseudorandom number generators for cryptographic applications. Technical report, 2010
8. K. Yang, D. Blaauw, D. Sylvester, An all-digital edge racing true random number generator robust against PVT variations. IEEE J. Solid State Circuits **51**(4), 1022–1031 (2016)
9. H. Jiang et al., A novel true random number generator based on a stochastic diffusive memristor. Nature Communications **8**(1), 882 (2017)
10. A. Fukushima et al., Spin dice: A scalable truly random number generator based on spintronics. Appl. Phys. Exp. **7**(8), 083001 (2014)
11. M.J. Collins, A.S. Clark, C. Xiong, E. Mägi, M.J. Steel, B.J. Eggleton, Random number generation from spontaneous Raman scattering. Appl. Phys. Lett. **107**(14), 141112 (2015)
12. P. Magnard, P. Kurpiers, B. Royer, T. Walter, J.-C. Besse, S. Gasparinetti, M. Pechal, J. Heinsoo, S. Storz, A. Blais, et al., Fast and unconditional all-microwave reset of a superconducting qubit. Phys. Rev. Lett. **121**(6), 060502 (2018)
13. M. Berta, Generating quantum randomness with Amazon Braket. https://aws.amazon.com/blogs/quantum-computing/generating-quantum-randomness-with-amazon-braket/. Accessed: 2022-02-20
14. AWS, Robust randomness generation on quantum processing units. https://github.com/aws/amazon-braket-examples/blob/main/examples/advanced_circuits_algorithms/Randomness/Randomness_Generation.ipynb. Accessed: 2022-02-20

15. C. Foreman, S. Wright, A. Edgington, M. Berta, F.J. Curchod, Practical randomness and privacy amplification. Preprint (2020). arXiv:2009.06551
16. N.D. Mermin, Quantum mysteries revisited. Am. J. Phys. **58**(8), 731–734 (1990)
17. D. Jones, Quantum-proof cryptography with Ironbridge, TKET and Amazon Braket. https://cambridgequantum.com/quantum-proof-cryptography-with-ironbridge-tket-and-amazon-braket/. Accessed: 2022-02-20
18. P. Rivero, QRAND: A multiprotocol and multiplatform quantum random number generation framework (2020)
19. J.E. Jacak, W.A. Jacak, W.A. Donderowicz, L. Jacak, Quantum random number generators with entanglement for public randomness testing. Scientific Reports **10**(1), 1–9 (2020)
20. A. Ash Saki, M. Alam, S. Ghosh, Experimental characterization, modeling, and analysis of crosstalk in a quantum computer. IEEE Trans. Quantum Eng. **1**, 1–6 (2020)
21. M. Piani, M. Mosca, B. Neill, Quantum random-number generators: Practical considerations and use cases. https://evolutionq.com/quantum-safe-publications/qrng-report-2021-evolutionQ.pdf. Accessed: 2022-02-20

Placement Algorithm of Superconducting Energy-Efficient Magnetic FPGA

Sagar Vayalapalli, Yi-Chen Chang, Naveen Katam, and Tsung-Yi Ho

1 Introduction

Superconducting Electronics (SCE) also known as quantum electronics can provide devices with some special properties like low noise, low loss, low power dissipation, less weight, high resolution, high speed, and high frequency. This is the result of the special attributes of Josephson junctions (JJs) [1] of SCE families, which enable fast switching ($\sim$1 ps) and low switching energy per bit ($\sim$10-9 J) at low temperatures. Josephson junctions are also the key component used in RSFQ circuits which the digital family of SCE and had shown some highest performance in recent years, with only a modest number of researchers worldwide [2, 3]. Due to the extraordinary properties of superconducting electronics, many researchers around are keenly enticed to work on this topic. Some researchers believe that superconducting electronics are the future of semiconductor companies in the coming few decades. Hence the availability of an SFQ-specific FPGA will be very useful soon.

FPGA plays a very vital role in semiconductor companies as they have many advantages over application-specific integrated circuits (ASICs). FPGAs can reduce the testing time and bypass the fabrication process resulting in reducing the design cycle effectively and efficiently [4]. One of the major features of FPGAs is that it is a re-programmable device and provides a flexible way of implementing digital circuits. It makes the hardware verification easier [5], once the fault has been

S. Vayalapalli (✉) · Y.-C. Chang · T.-Y. Ho
National Tsing Hua University, Hsinchu, Taiwan
e-mail: sagar@gapp.nthu.edu.tw

N. Katam
Seeqc Inc., Elmsford, NY, USA
e-mail: tyho@cs.nthu.edu.tw

encountered, it is unlikely to be fixed on ASICs whereas it can be fixed and re-programmed on FPGAs. Hence, this lowers down the cost of the design flow.

In recent years, there is a rapid growth on SFQ circuits. SFQ and CMOS technology differ in some ways which can be summarized as follow [6]:

- Superconducting devices require Cryogenic temperature to operate.
- Size of basic gates in SFQ circuits is much larger than CMOS circuits, i.e., an AND gate of SFQ circuits is about 600 times larger than 45 nm CMOS AND gate.
- In current SFQ technology, there are only 3 layers for both clock and signal routing, i.e., only limited layers are available in SQF technology at present, unlike CMOS technology.
- In SFQ circuits nearly all the cells need a clock except the splitters, unlike CMOS circuits where only 15% of cells need clock signal.
- In SFQ designs, DFFs are inserted to the input netlist so as to path balance it. This increases the number of cells in the design, eventually increasing the total area.
- A three-terminal device like MOSFET in CMOS technology is not yet available in SCE technology, therefore CMOS-based FPGA cannot be directly considered to develop RSFQ FPGA. Due to this, SFQ technologies does not support the major benefits of CMOS technology and bidirectional wires for programming routing in RSFQ FPGA turn out to be more difficult. Hence, all the wires in RSFQ FPGA are unidirectional making the programmable routing network more complex.

The organization of the chapter is as follows. The related works are shown in Sect. 2. Section 3 discusses the architecture of RSFQ FPGA with a brief explanation of each block of the FPGA. Section 4 analyzes the CAD problem for RSFQ FPGA with a detailed explanation of design constraints followed by the proposed method in Sect. 5. The placement algorithm consists of three main attributes, namely: cell grouping, global placement, a detailed placement that produces the best legal placement solution complying with all the RSFQ FPGA design constraints.

2 Related Work

The first Superconducting FPGA was proposed in 2007 [7] which was relied on the implementation of Superconducting Quantum Interference Devices (SQUIDs) and Non-Destructive Read-Out circuit (NDROs) controlled by dc bias to program LUT (Look Up Table) and routing for a logic block in FPGA fabric. The cell area of SQUID and NDRO is very large around $(40 \times 60)\,\mu m^2$ and requires bias current of $1500\,\mu A$.

We proposed a first superconducting energy-efficient FPGA in 2018 [8], a complete FPGA fabric architecture with all the necessary blocks and circuits. It is based on programmable dc-bias controlled by magnetic JJs (MJJs), because MJJ

Table 1 Comparison of NDRO with MJJ		NDRO gate with (I/O) JTL	MJJ
	Area μm²	40 × 60	2 × 2
	Bias current μA	Not less than 1500	No bias current

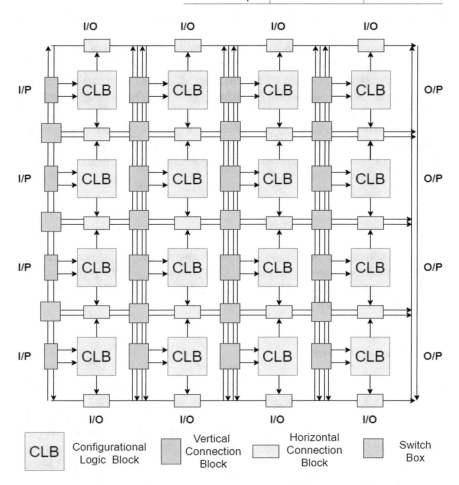

Fig. 1 RSFQ FPGA Architecture. Island-style adaption in which a CLB gets input through VCB from routing network and the output from CLB goes to routing network through HCB with bidirectional and unidirectional data flow in vertical and horizontal directions, respectively. I/P: Input, O/P: Output, I/O: Input/Output

has very little area typically around $(2 \times 2)\,\mu m^2$, high switching speed, and no bias current is needed. This makes RSFQ FPGA operate at a very high speed with low power dissipation (see Table 1). Hence, it makes it as a first superconducting energy-efficient magnetic FPGA. The detailed explanation of RSFQ FPGA architecture with all the necessary blocks can be seen in Sect. 3.

3 Superconducting FPGA: Architecture and CAD

3.1 RSFQ FPGA Architecture

Semiconductor companies like Altera [9] and Xilinx [10] have provided different types of commercial FPGA Architectures, however, all these architectures contain CLB (Configuration Logic Block), Switch Boxes, and Routing Channel.

RSFQ FPGA architecture is based on Island-Style Architecture also known as hierarchical architecture where CLBs are arranged in an array of two-dimensional logic modules on a grid connected by a routing network. It is known as island-style architecture because the 2-D grid is like an open sea where each CLB looks like an island with interconnects. RSFQ FPGA includes CLBs, VCBs (vertical Connection Box), HCBs (Horizontal Connection Box), Switch Boxes, I/O pins, and routing networks. The detailed explanation of RSFQ FPGA architecture can be seen in Fig. 1 [8].

3.1.1 Configuration Logic Block

The main component of FPGAs is logic resources that implement and store the functionalities of the target circuit. CLB executes complex logic functions, implements memory functions, and synchronizes code on FPGA when linked together by routing resources.

3.1.2 Routing Network

Figure 1 shows that the horizontal data flow in RSFQ FPGA is only unidirectional, i.e., from left to right containing two horizontal unidirectional data lines and the vertical data flow is bidirectional provided that there are two separate lines for "top to down" and "down to top," respectively. Thus, the input and output ports are located on the left and right sides of the FPGA grid. Both the input/output ports are located on the top and bottom sides of FPGA, unlike a typical CMOS-based FPGA, where all the input and output ports are located on all sides of the FPGA grid [11].

3.1.3 Switch Box

The Wilton switch is modified in a way to configure it to the unidirectional routing architecture of SFQ FPGA and scalable for a large number of routing channels. The switch box implementation is presented in Fig. 2a [8] and it comprises 3-to-1 and 2-to-1 mergers (see Fig. 2b,c), three-way and two-way splitters (see Fig. 2d,e). L1

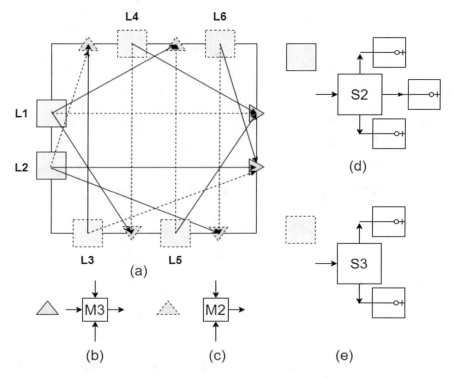

Fig. 2 (**a**) Switch box implementation. (**b**) Three signal merger. (**c**) Two signal merger. (**d**) Three-way splitter (S3) with attached switches at outputs. (**e**) Two-way splitter (S2) with switches

and L2 are horizontal lines from left to right. L3, L6, and L4, L5 are vertical lines from bottom to top and top to bottom, respectively (Fig. 3).

3.1.4 Connection Blocks(CBs)

In the RSFQ FPGA, the HCB and VCB are responsible for the connection between CLBs and the routing network. In other words, VCBs and HCBs are the interfaces between CLBs and the routing network. VCBs are responsible for taking inputs to CLBs from the routing network and output of CLBs are taken to the routing network through HCBs. It is presented in Fig. 4 [8].

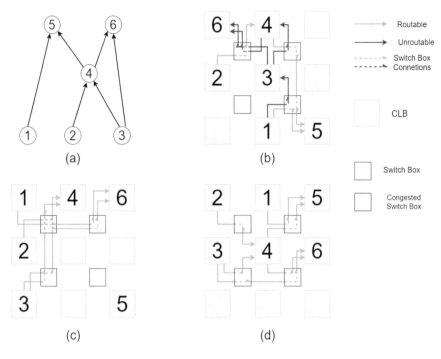

Fig. 3 RSFQ FPGA placement and routing constraints. (**a**) Simple circuit netlist. (**b**) Routability and unroutability of RSFQ FPGA. (**c**) Switch box congestion. (**d**) Best placement

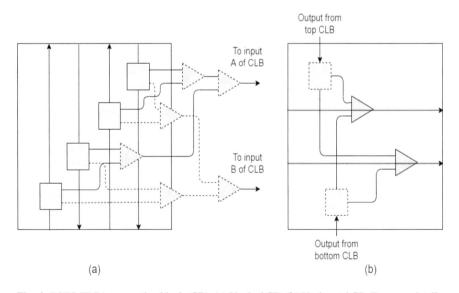

Fig. 4 RSFQ FPGA connection block (CB). (**a**) Vertical CB. (**b**) Horizontal CB. For more details, the reader is referred to [8]

4   RSFQ FPGA Placement

4.1   Problem Formulation

Given an RSFQ FPGA with heterogeneous Logic Architecture and a circuit netlist with logic type specification, the placement problem is to assign all the placeable modules to their legal locations (sites) such that some predefined objective function (e.g., total wirelength and placement area) is optimized and minimize the congestion on switch boxes. Each module is square and of the same size.

4.2   SFQ-Specific FPGA Placement Constraints

Placement in RSFQ FPGA is completely different in contrast with placement in commercial CMOS FPGA. This is because SFQ circuits have new or different constraints with respect to CMOS circuits. An illustrated example of RSFQ FPFA placement and routing with their constraints are shown in Fig. 3. All the design constraints are listed below:

- All the data flow lines are unidirectional.
- Data flow is only in a forward direction horizontally, i.e., the left to right direction of the FPGA grid. Data cannot flow in a backward direction horizontally in any case. Hence, feedback circuits cannot be placed and routed in the RSFQ FPGA.
- Cells placed in the same column of the FPGA grid cannot communicate with each other, i.e., they cannot be connected in any way.
- A limited number of I/O pins: Input (I/P) and Output (O/P) ports are located on the left and right sides of the FPGA grid, respectively. Both Input/Output (I/O) ports are located on the top and bottom sides of FPGA, unlike a typical CMOS FPGA where Input/Output (I/O) ports are located all around the grid.
- Any two random cells cannot be interchanged in RSFQ FPGA, unlike a typical CMOS FPGA.
- No site is assigned more than one cell, i.e., no overlapping of cells is allowed.

5   Proposed Method

A great variety of placement algorithms for FPGA like simulated annealing, partition-based method, and analytical placement were proposed and these algorithms tend to produce high-quality placement results. However, our method is based on Simulated Annealing as it obtains ultra-fast placement with good quality and it is easy to implement [11].

5.1   Cost Function

The cost function of RSFQ FPGA design might not be the same as those of classical CMOS FPGAs that are available in the market. But, wirelength is often one of the most basic metric for consideration because smaller wirelength typically leads to less wire resource consumption, less congestion on switch boxes and smaller delay [12]. The distance between two cells is measured from the center of one cell to the other.

The cost function C(f) can be defined as the summation of total wirelength between all the connected nodes, which can be represented mathematically as below:

$$C(F) = \sum_{i=1}^{n} (X_{i+1} + Y_{i+1}) - (X_i + Y_i)$$

where (X_i, Y_i) are the indexes of Cell i in the grid.

5.2   Methodology

Our placement algorithm consists of three main stages: 1. Cell Grouping; 2. Global Placement; 3. Detailed Placement; (See Figs. 5 and 6).

5.2.1   Cell Grouping

In this stage, the main idea is to group the cells which are at the same levels as shown in Fig. 7. The level in a graph can be determined by partitioning the vertices into subsets that have the same distance from a given root vertex. Initially, the input circuit netlist is read and the netlist is transformed into a Graph $G(E, V)$, where E is set of all Edges and V is set of all vertices. Then, it is checked whether the netlist contains any feedback loop (cycle). The placement is said unroutable if there is a cycle in the netlist. Hence, it is necessary to check whether the input circuit is routable or unroutable in the beginning. Then, the netlist is converted into a smaller netlist containing supercells, where each supercell is a set of cells. The supercells are determined based on the depth of each vertex in the graph. Depth of a vertex is the number of branches or edges E in the path from root to the vertex V in graph G (see Algorithm 1).

This algorithm gets an input of graph $G(E,V)$ and returns a two-dimensional vector, namely *supercells*, containing all the supercells of the graph G. Algorithm 1 starts by initiating an empty *queue*, vector *add_list* and a two-dimensional vector *Supercells*. Initially, the input vertices are pushed into *queue* and marked as first *supercell*. For each time, when the *queue* is not empty and for each vertex *v* in Q,

Fig. 5 Placement algorithm
flow chart

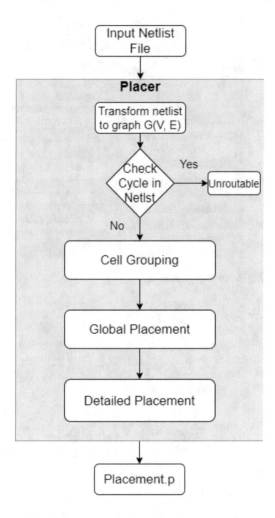

we obtain its adjacent vertex x in the graph G and push them into *queue*. Now, pop all the previous level nodes from *queue* and sort the *queue*. Then, for each vertex v in *queue*, check whether there is a path between vertex v to other vertices present in *queue*. If there is a path between them, push them into *Ignore_list*, else push them into *Add_list*. Repeat this process for each vertex in *queue*. At the end of the iteration, the elements present in *Add_list* are the vertices of the next level. Hence, push them into two-dimensional vector *supercells* marking them as a next level cells. Now, to get ready for the next iteration, clear *queue* and push all the elements of *Add_list* into *queue* and finally clear the vector *Add_list* to save the new elements in the next iteration. This process ends when we get the last level vertices into *supercells* and the *queue* is empty.

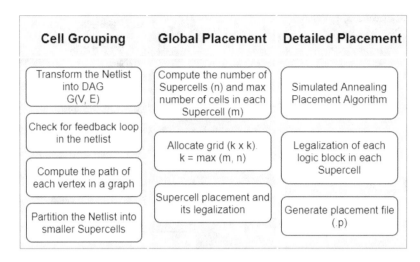

Fig. 6 Overall placement

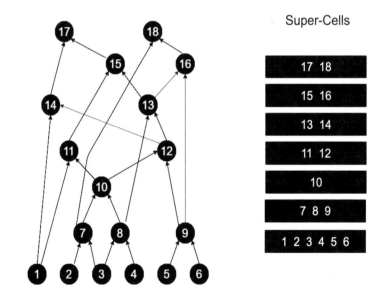

Fig. 7 Cell grouping

5.2.2 Global Placement

Once all the supercells are generated and the cells are grouped, the placement of these supercells initiates by creating a grid with some initial area $(m \times n)$ where m denotes the number of rows and n denotes the number of columns of the grid as shown in Fig. 8. The area for global placement is allocated by computing the total number of supercells generated, i.e., m and the maximum number of cells in a

---

Algorithm 1 Cell grouping

1: vector<vector<int>> Supercella
2: vector<int> add_list
3: Queue<int> Q
4: **for** Each vertex v in Graph G **do**
5:     **if** no in_degree **then**
6:         Q.push_back(v) //input nodes
7:     **end if**
8: **end for**
9: Supercells.push_back(Q)
10: **while** !Q.empty() **do**
11:     **for** Each vertex v in Q **do**
12:         **for** Each adjacent vertex x of v **do**
13:             Q.push_back(x)
14:         **end for**
15:         Q.pop()
16:     **end for**
17:     **for** Each vertex v in Q **do**
18:         **if** any other vertices of Q lies in the path of v **then**
19:             Continue
20:         **else**
21:             Add_list.push_back(x)
22:         **end if**
23:     **end for**
24:     Supercells.push_back(Add_list)
25:     Q.clear()
26:     **for** Each vertex v in Add_list **do**
27:         Q.push_back(v)
28:     **end for**
29:     Add_list.clear()
30: **end while**
31: **return** Supercells

---

supercell, i.e., n. The original circuit netlist is transformed into a smaller netlist containing all supercells [6]. In this phase, the legal locations of supercells are obtained and placed using the global placement algorithm. The legal location of each supercell is determined considering all the constraints (Sect. 4.2) in the design. The global placement algorithm places each supercell on each column of the grid such that nth supercell is placed on nth column of the grid. In other words, the first supercell is placed on the first column, the second supercell is placed on the second column, and so on.

5.2.3 Detailed Placement

Once all the supercells are placed and the global placement algorithm is ended, a detailed placement algorithm (see Algorithm 2) starts to refine the solution quality. Detailed Placement focuses on cells inside each supercell to obtain their best loca-

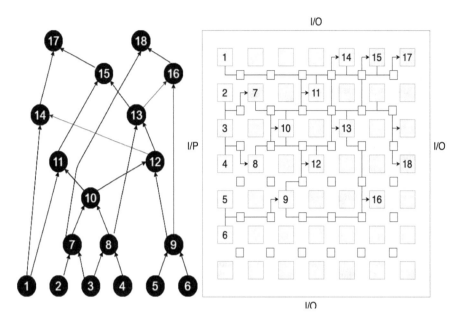

Fig. 8 Global placement. Initial placement area = 6 rows × 7 columns = 42. Total initial wire-length = 53

Algorithm 2 Detailed placement

1: Set Initial_temperature (T_0)
2: **while** $T_0 > 0.01$ **do**
3:     **for** Each column (col1) in a grid **do**
4:         **for** N trails **do**
5:             Select any random cell (v1) from col
6:             Select any random column (col2) in grid
7:             Select any random cell (v2) from col1
8:             **if** legalize(v1, col1, v2, col1) **then**
9:                 Continue
10:             **end if**
11:             Swap v1 and v2
12:             Compute new_cost
13:             δ = new_cost - previous_cost
14:             **if** $\delta < 0$ **then**
15:                 Accept new placement
16:                 previous_cost = new_cost
17:             **else**
18:                 discard new placement
19:             **end if**
20:         **end for**
21:     **end for**
22:         $T_0 \times 0.95$
23: **end while**

Algorithm 3 Legalize(v1, col1, v2, col1)

1: **if** v1, v2 = NULL **then**
2:       **return** True
3: **end if**
4: **if** col1 = col2 **then**
5:       **return** False
6: **end if**
7: **if** (v1 linked to col2) ‖ (v2 linked col1) **then**
8:       **return** True
9: **end if**
10: **return** False

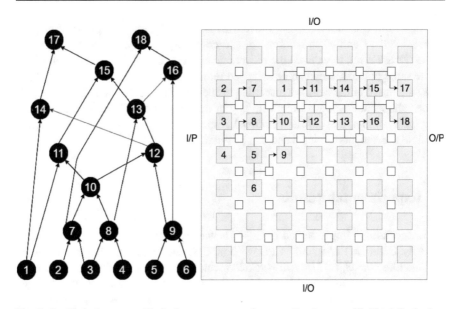

Fig. 9 Detailed placement. Final placement area = 4 rows × 7 columns = 28. Total final wire-length = 32

tion on the grid as shown in Fig. 9. Our approach is based on a simulated annealing algorithm as it promises to provide the placement results. The technique of detailed placement is based on moving cells to some empty locations and swapping locations of two cells in the grid followed by legalization of the movement. Note that, in this process, no overlap of two cells is allowed. After each iteration, a new placement is generated and it is evaluated by a predefined cost function to determine whether this new placement is accepted or discarded. It is always accepted if the new cost function is smaller than the previous cost function else we discard the move and place back the cells in their original locations. Thus we refine the solution at each iteration till we obtain the best solution.

Algorithm 2 starts by scheduling high *initial temperature* at the beginning and as the name suggests *Annealing* which means gradually decreasing it to some minimum temperature over each iteration. During this process, for each column, *col* (starting from column 1 to column n), there might be some set of feasible movements (say 100) of cells in the grid. Each time, a cell *v1* can be moved to any empty location in the grid or swapped with another cell *v2* from any random column *col1*, satisfying the legalization of the movement. If the movement of the cells does not disregard any rule, then the *new cost* of this solution is computed. If the *new cost* is less than the *previous cost* then we accept the new placement making it a new original placement. The legalization of the movement can be seen in *Algorithm 3*. This process continues for each column of the grid and ends when the temperature is lower down to some *cooling temperature* providing the optimized placement solution.

Algorithm 3 examines if the movement is legal. It gets an input of randomly selected cell *v1* and *v2* in column *col* and *col1*, respectively. It returns a Boolean value *(True/False)*, to validate the movement as legal or illegal. If cell *v1* and *v2* are in the same column, they can be swapped returning *False*, as it is legal movement. The movement of the cells are discarded or illegal: *(1)* if cell *v1* and *v2* are zero, *(2)* if the cells *v1* and *v2* have links with any of the cells of each other's columns, i.e., *col1 (v2's column)* and *col (v1's column)*, respectively. Apart from that, all the other possible movements are legal and can be interchanged.

6   Experimental Results

The Algorithm is implemented in C++. It was executed on system configuration: Intel(R) Core(TM) i7-9700K CPU @ 3.60 GHz and tested on large benchmark circuits like *C1908*, *C5315*, *ALU4*, *C6288* which can be seen in Table 2. In Table 2 the *initial placement* is nothing but a global placement solution which is a legal placement and it has been improved in the final placement. The *final placement* is

Table 2 RSFQ FPGA placer

Benchmark circuit	No. of logic blocks	Initial placement Cost	Area	Final placement Cost	Area	% Reduction Cost	Area	Execution time (s)
C1908	652	19,395	46×63 2898	7509	47×41 1927	61.28	33.50	98.901
C5315	2315	254,403	239×92 21,988	63,101	118×91 10,738	75.19	51.16	310.008
C6288	2481	607,134	554×66 36,564	198,865	265×81 21,465	67.24	41.29	189.368
ALU4	2658	726,264	679×74 50,246	230,247	385×92 35,420	68.29	29.50	201.495

This tool was executive on System Configuration: Intel(R) Core(TM) i7-9700K CPU @ 3.60 GHz

Table 3 Baseline placement approach for RSFQ FPGA

Benchmark circuit	No. of logic blocks	Initial placement		Final placement		% Reduction		Execution time (s)
		Cost	Area	Cost	Area	Cost	Area	
C1908	652	19,395	46 × 63 2898	13,532	59 × 41 2419	30.22	16.50	66.10
C5315	2315	254,403	239 × 92 21,988	162,251	178 × 91 16,198	36.22	26.33	146.10
C6288	2481	607,134	554 × 66 36,564	312,863	348 × 81 28,188	48.46	22.90	79.56
ALU4	2658	726,264	679 × 74 50,246	450,721	469 × 92 43,148	37.93	14.12	120.51

This tool was executive on System Configuration: Intel(R) Core(TM) i7-9700K CPU @ 3.60 GHz

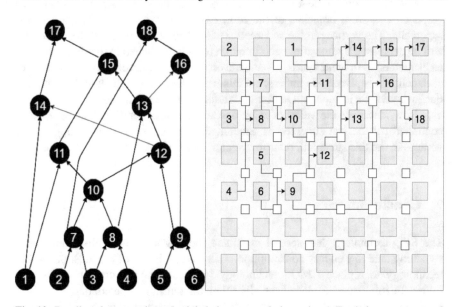

Fig. 10 Baseline placement: it needs global placement solution as input. Total placement area = 5 rows × 7 columns = 35. Total wirelength = 45

the optimization of *initial placement* solution which includes discovering the best location of each cell in the grid followed by its legalization. The results show that our algorithm was capable of reducing the average cost and placement area by 68% and 39%, respectively. The placement results of RSFQ FPGA are also compared with the Baseline solution shown in Table 3. Although, the baseline solution fulfils all the RSQF constraints, it cannot be considered as a perfect placement solution because it requires more wirelength and area to produce final placement solution as shown in Fig. 10.

Algorithm 4 Baseline placement algorithm

1: vector<int> supercell (input from global placement results)
2: number of supercells = n (input from Algorithm 1)
3: **for** each vertex (v) in supercell (n) **do**
4:     **for** each adj. vertex (x) of v **do**
5:         move x into supercell n-1
6:     **end for**
7: **end for**
8: n = n - 1
9: **return** supercell

The baseline algorithm takes the global placement results, a graph G, a vector *supercell*, and the number of supercells n as input and returns the updated vector *supercell* as final placement results. The algorithm starts by reserve traversing the graph G, i.e., from branches to the root, based on their present location in the supercells. The vertices in the supercell n will be traversed first and the value of n will be decreased by one after each iteration until it is reached to the first supercell, i.e., supercell 1. In other words, reverse traversing from supercell n to supercell 1, for each cell in a supercell. To understand this better, let us take a look at Algorithm 4. For each vertex v in supercell n, obtain its adjacent vertex x and move it to supercell *n-1*. Now, decrease the value of n by one for the next iteration. This process is continued till the last supercell is encountered and stops after traversing all the vertices in the this supercell, returning vector *supercell* as a final placement results.

7   Conclusion

The SFQ technology has special attributes such as low power dissipation, fast switching, and low energy consumption. Therefore, it is one of the most promising technology to replace CMOS technology in semiconductor companies in near future. In this paper, a placement algorithm for the first superconducting FPGA is proposed and developed. Its virtue of proposed method is collated with state-of -art-techniques. This paper also promotes the significant advancement of RSFQ FPGA. The proposed approach can minimize the congestion on switch boxes by reducing wire consumption and placement area by 68% and 39% of initial to final placement, respectively. It also talks about the basic difference between CMOS and SCE technologies.

The RSFQ placer results were also compared to the baseline placement algorithm. Both the algorithms were executed with the same netlist files and system configuration. The baseline algorithm takes 35% of less execution time but takes 55% of more wirelength and 51% of more area to produce final placement. Hence, the RSFQ placer is more cost-efficient that promises to produce the best placement results with minimum wirelength and area.

References

1. D. Averin, A. Bardas, ac Josephson effect in a single quantum channel. Phys. Rev. Lett. **75**(9), 1831 (1995)
2. D.K. Brock, RSFQ technology: circuits and systems. Int. J. High Speed Electron. Syst. **11**(01), 307–362 (2001)
3. K.K. Likharev, Rapid single-flux-quantum logic, in *The New Superconducting Electronics* (Springer, Dordrecht, 1993), pp. 423–452
4. J. Monteiro, R. Van Leuken (eds.) *Integrated Circuit and System Design: Power and Timing Modeling, Optimization and Simulation: 19th International Workshop*, PATMOS 2009, Delft, September 9–11, 2009. Revised Selected Papers, vol. 5953 (Springer Science & Business Media, Berlin, 2010)
5. A.G. Ye, Using the minimum set of input combinations to minimize the area of local routing networks in logic clusters containing logically equivalent I/Os in FPGAs. IEEE Trans. Very Large Scale Integr. Syst. **18**(1), 95–107 (2009)
6. S.N. Shahsavani, A. Shafaei, M. Pedram, A placement algorithm for superconducting logic circuits based on cell grouping and super-cell placement, in *2018 Design, Automation & Test in Europe Conference & Exhibition (DATE)* (IEEE, Piscataway, 2018), pp. 1465–1468
7. C.J. Fourie, H. van Heerden, An RSFQ superconductive programmable gate array. IEEE Trans. Appl. Supercond. **17**(2), 538–541 (2007)
8. N.K. Katam, O.A. Mukhanov, M. Pedram, Superconducting magnetic field programmable gate array. IEEE Trans. Appl. Supercond. **28**(2), 1–12 (2018)
9. D. Singh, Implementing FPGA design with the OpenCL standard. Altera whitepaper 1 (2011)
10. A. Cosoroaba, F. Rivoallon, Achieving higher system performance with the Virtex-5 family of FPGAs. White Paper: Virtex-5 Family of FPGAs, Xilinx WP245 (v1. 1.1) (2006)
11. U. Farooq, Z. Marrakchi, H. Mehrez, FPGA architectures: an overview, in *Tree-Based Heterogeneous FPGA Architectures* (Springer, Berlin, 2012), pp. 7–48
12. S.-C. Chen, Y.-W. Chang, FPGA placement and routing, in *2017 IEEE/ACM International Conference on Computer-Aided Design (ICCAD)* (IEEE, Piscataway, 2017), pp. 914–921
13. V. Betz, J. Rose, VPR: a new packing, placement and routing tool for FPGA research, in *International Workshop on Field Programmable Logic and Applications* (Springer, Berlin, 1997), pp. 213–222

Margin Optimization of Single Flux Quantum Logic Cells

Mustafa Altay Karamuftuoglu, Soheil Nazar Shahsavani, and Massoud Pedram

1   Introduction

Developments in superconducting logic cell families create alternative choices for the next generation integrated cell and system designs due to low power dissipation and high speed operation characteristics [1]. Optimizing the parameter values to be robust against the variations in the fabrication process is a crucial task for Single Flux Quantum (SFQ) logic families. Margin calculation and yield analysis are two of the most common techniques for evaluating the robustness of SFQ logic cells against process variations. Consequently, when optimizing logic cells, maximizing operating margins is a primary objective in designing robust logic cells with high tolerances to variations in the fabrication process [2]. Note that an increase in the critical margin of a cell results in a higher parametric yield, which in turn corresponds to improved robustness of a cell to the process variations.

The most common objective function in margin optimization is to maximize the summation of lower and upper critical margins, i.e., maximizing the smallest margin among all the parameters [2]. However, such an objective function can be a poor indicator of the robustness of a logic cell as lower and upper bound (LB & UB) margins may not necessarily be symmetrical. Moreover, SFQ cells have different characteristics when interfacing to different fan-in and fan-out cells. Therefore, the optimization process should be repeated for all different combinations of fan-in and fan-out cells, which greatly complicates the optimization process. For instance, loading a target cell with a low upper bound margin with another cell having a small lower bound margin may significantly alter the characteristics of the target cell and lead to poor parametric yield values. Therefore, developing optimization

M. A. Karamuftuoglu (✉) · S. Nazar Shahsavani · M. Pedram
University of Southern California, Los Angeles, CA, USA
e-mail: karamuft@usc.edu; nazarsha@usc.edu; pedram@usc.edu

© The Author(s), under exclusive license to Springer Nature Switzerland AG 2023   105
R. O. Topaloglu (ed.), *Design Automation of Quantum Computers*,
https://doi.org/10.1007/978-3-031-15699-1_6

algorithms that account for such mechanism when optimizing a target cell is of high importance. Details of objective functions are provided in Sect. 2.2.

Many optimization tools (e.g., MALT [3], COWBoy [4], xopt [5], SCOPE [6] and an optimizer utilizing PSO [7]) have been developed to find high-quality design solutions for SFQ cells. In the following, we give a quick overview of the algorithms that have been proposed for SFQ cell optimizations. More detailed explanations will be provided in Sect. 2.3.

Herr and Feldman (1995) proposed the use of Inscribed Hyperspheres Method (IHM) for Rapid Single Flux Quantum (RSFQ) circuit optimization [8]. The algorithm models the range of the parameter set as a sphere. In an attempt to achieve better yield values, Harnisch et al (1997) applied Centers-of-Gravity Method (CGM) for design centering on SFQ cells [5]. Their proposed algorithm defines acceptability regions and uses Monte Carlo (MC) simulations within these regions. Fourie and Perold (2003) investigated the optimization of SFQ cells using Genetic Algorithms (GA) [9]. In this work, the authors map real-valued parameters of SFQ cells into binary substrings in the genomes, i.e. strings, chromosome. Tukel et al (2012) and Karamuftuoglu et al (2016) investigated the application of Particle Swarm Optimization (PSO) to the SFQ cell optimization problem [7, 10]. In this technique, each particle represents a set of parameter values in an attempt to find a location that has the best parameter set for the desired objective function.

Since our proposed optimizations for SFQ cell build on the PSO algorithm, we provide a quick review of the PSO algorithm next. PSO simulates the social behavior of swarms such as bird flocks (or fish schools) in nature where particles corresponds to a bird (or fish) to find the best environment. The area outside of the search space is considered as the forbidden area, where the birds (particles) are not allowed to fly over even if the environment has a better food offering. In the standard PSO approach, particles converge into a single point that optimizes the objective function and it suffers from premature convergence, i.e., getting trapped in local optima. Additionally, the exploration process within the search space mainly focuses on global search. Thus, the results are strongly dependent on each particle's movement towards the elite particle (which identifies a point of attraction for all the particles) [11–13].

This chapter presents a hybrid cell optimization methodology (ANPS-FW) employing the Automatic Niching Particle Swarm Optimization (ANPSO) [14] and Fireworks Algorithm (FWA) [15]. ANPSO, which is a modified version of PSO that extends the unimodal PSO to efficiently locate multiple optimal solutions in multimodal problems, is a powerful technique for increasing the chances of finding a better solution during the global search. The inertia weight is one of PSO parameters to bring about a balance between the exploration and exploitation characteristics of PSO. ANPSO relies on an adaptive inertia weight to make a sensitive search in the late iterations [16]. FWA uses an effective local search strategy around each global optimum obtained in each iteration of the ANPSO. By combining ANPSO with FWA, the proposed method can provide a better local search around each particle so as to maximize the chances of finding a better result.

ANPSO is a population based algorithm for stochastic search in a multi-dimensional space, which optimizes an objective function by trying to iteratively improve a candidate solution with respect to a given figure of merit (objective function). In the context of SFQ cell optimization, the search space is the set of all possible parameter values. Particles are created and added to the system, and eventually die after a certain amount of time. Each particle can be assigned its own range of motion, and behavior including location, size, velocity, etc. ANPSO places particles, i.e, a population of candidate solutions, and creates subswarms within the search space. Each group of particles is considered a subswarm since the whole population is divided into multiple batches. Each particle seeks to find the best location defined by the objective function. A group of particles is assigned to a closest best known solution, which is in turn selected by the Roulette Wheel Selection (RWS) algorithm [17]. Experimental results show that the RWS algorithm for particle selections outperforms the conventional Rank Selection (RS) algorithm. If RS picks a local maximum before other possible maximum points are found, the algorithm will continue with a local search of the initially selected point only. However, in the same scenario for RWS, other possible convergence points of the population still have a chance to be chosen, which allows the particles to move into different local regions, thereby, avoid premature convergence to a local maximum and facilitate finding a better critical margin for SFQ cells.

FWA is a swarm intelligence algorithm based on parallel diffuse optimization to simulate the fireworks explosion phenomenon, which can achieve a good balance between global exploration and local searching by means of adjusting the explosion parameters. More precisely, in FWA first a fixed number of (S_M) locations in the search space are chosen to start fireworks. The volume and the direction of the fireworks explosion have adaptive characteristics corresponding to the quality of a firework [15]. Then, a set of (S_N) random sparks at each firework location are generated, and locations of the sparks are calculated. Next, the quality of each spark location is evaluated and the best solutions (locations) from the whole sets of fireworks and sparks are selected as the firework locations for the next generation. The selection is also driven by the desire to maintain the diversity of the new set of fireworks locations. The process is iteratively done to improve the quality of the solution. In each iteration, a certain number of locations within the best locations are selected based on a probability value obtained from distance information such as Manhattan distance, Euclidean distance, or Angle-based distance [18].

The intrinsic combination of ANPSO and Fireworks algorithms is performed in such a way that one can have better balance between exploration vs. exploitation. PSO is already proven to work for SFQ cell optimization as in [7] whereas the Fireworks algorithm has been successfully used by many researchers. Therefore, we have opted to improve the optimization solutions for SFQ cells by combining two swarm optimization techniques which was never applied before.

The combination of ANPSO and FWA is achieved by merging the next location calculation process of ANPSO with the firework explosion feature of FWA. The particle locations determine where the fireworks start. The quality of fireworks depends on the number of sparks generated by the explosion. In this work, particle locations and the number of generated sparks are controlled and corrected by using the Damping Wall (DW) technique [19]. DW decreases the spark's excessive velocity and redirects the sparks when the location surpasses the search space boundaries. In other words, the global search is carried out by particles and the local search is performed by firework sparks. By changing the number of particles or each firework's spark reach, the margin optimization algorithm can expand either the global search or the local search around the target SFQ parameter values in the search space. During the optimization process, multi-threading is employed to speed up the optimization process, where each thread is assigned to a single particle or a firework.

For calculating the objective function, the conventional margin analysis is performed where every parameter is scanned through its own boundary limits while the other parameters remain the same at their nominal values. The margin calculation is completed by using binary search on every parameter within given parameter boundaries [2]. For the critical margin range calculation in this study, two factors are considered: the square root of the product of the lower and upper margin values and the sum of lower and upper margin values. The square root value carries the information of how well-centered the margin range is, whereas the sum provides information of how much a cell is tolerant to the process variations. Moreover, a similar approach can be applied to complex cells for simultaneous optimization of parameter margins when dealing with different fan-in and fan-out cells. This will facilitate generating complex cells by combining multiple cells from a cell library. During the optimization process, it is also possible to group similar parameters for cells where it is intended to optimize cells with a symmetrical structure such as Superconducting Quantum Interference Device (SQUID) [20].

2   Background Knowledge

This section begins with a brief explanation of SFQ logic with an example of D Flip-Flop. In the following Sects. 2.2 and 2.3, the utilized objective functions to evaluate the SFQ cell performance and applied optimization methods for SFQ cells are discussed.

2.1   SFQ Logic

SFQ logic cells and interconnects consist of Josephson Junctions (JJs), inductors, and resistors. These circuits perform pulse-based operations where logic 1 corresponds to an SFQ pulse appearance and logic 0 corresponds to the absence of a pulse. The correct functionality of these circuits is determined by observing the pin-to-pin pulse arrival time and propagation delays. When an SFQ pulse does not appear within the desired time interval, the cell functionality is considered to be

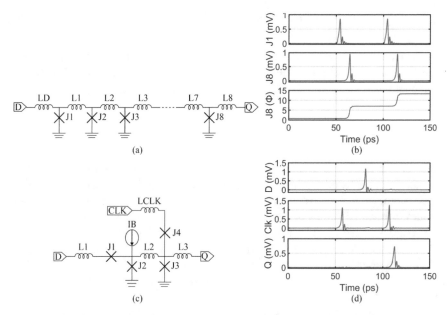

Fig. 1 (a) JTL schematic. (b) JTL simulation result. (c) DFF schematic. (d) DFF simulation result

incorrect. Depending on the cell type and its design, SFQ pulses can be transferred or stored by utilizing DC SQUID structure. A common interconnect and the simplest design used in RSFQ is Josephson Transmission Line (JTL) shown in Fig. 1a. JTL allows an incoming SFQ pulse to propagate to the output. It contains parallel JJs with series inductors in between. If the input pulse coming from A excites the first junction J1, the junction will have $2\text{-}\pi$ phase leap, forming supercurrent which will circulate the superconducting loop J1-L1-J2. When this supercurrent is added to the bias current of J2 and if the summation surpasses the critical current of J2, an SFQ pulse will be generated. The same process is repeated for the subsequent superconducting loops until the pulse is transmitted to the output node B as shown in Fig. 1b.

To store an SFQ pulse, D Flip-Flop (DFF) is built around a DC SQUID J2-L2-J3 where a quantizing SFQ loop is formed as shown in Fig. 1c. When an SFQ pulse is provided from input D given in Fig. 1d, the state of the quantizing loop becomes 1 and the DC-current flows in the clockwise direction while biasing J2 far from its critical current. If another SFQ pulse applied to D, the escape junction J1 switches, and the DFF remains in state 1. The DFF state can be read by applying a pulse from clk. If the DFF is in state 1, junction J3 will flip, releasing the stored flux quantum. If the DFF state is 0, junction J4 will flip since it is closer to its critical current value than J3.

2.2   Objective Functions for SFQ Design

The performance of SFQ cells is determined by evaluating the following objective functions. Each function requires a different computational effort to deal with SFQ cell components as will be explained next.

2.2.1   Margin

One of the most important metrics to evaluate the robustness of logic cells to variations in the fabrication process is the (critical) margin of a logic cell. Margins of each parameter in a logic cell are defined as the ratio of the minimum and maximum values to nominal values of cell parameters (e.g., bias current, Josephson junction (JJ) critical current, and inductance values) for which the cell functions correctly. These bounds are obtained by changing the nominal value of each parameter while the other parameter values are fixed. The correct functionality of each cell can be determined by observing the timing as well as the output values of a logic cell with respect to an input vector.

The critical margin is calculated as the summation of the maximum of lower margins and the minimum of upper margins of all parameter values. For example, if DFF cell parameter J2 has +5%/−10% upper and lower margins whereas another parameter J3 has +15%/−5%, the critical margin will be defined as the intersection region of both parameter margin regions, i.e., the critical margin with respect to J2 and J3 is +5%/−5%. Any improvement on the critical margin means every parameter are more tolerant to the variations in the fabrication process. When optimizing the margins or the critical margin, the parametric yield of logic cells is also improved as the optimized cell is more tolerant to process variations.

To reduce the dimensions of search space, a new margin calculation algorithm is reported by Shahsavani and Pedram [21]. Conventional margin analysis is performed by single parameter change and finding the overlapping regions of all parameters. However, the presented technique clusters the parameters into hyper-parameters to calculate accurate margin ranges. This allows them to perform multi-dimensional binary search and to find feasible parameter regions which will also help to quickly estimate the parametric yield.

2.2.2   Yield

Another employed technique for performance and reliability optimization is parametric yield analysis. Yield is a measure which captures the percentage of Monte Carlo (MC) simulations that maintain the correct cell functionality under specified conditions. These conditions can be the number of pulses or pulse timings observed at desired nodes in SFQ cells. For this analysis, independent normal distributions are assumed for different parameter types in the target cell, and a randomization process is performed to scatter parameter values around their nominal values subject to their respective standard distributions. Yield (Y) is reported as a percentage that shows the ratio of the samples that pass the desired conditions for correct output (N_p) to the total number of samples (N), that is,

$$Y = \frac{N_p}{N}$$

$$N_p = \sum_{i=1}^{N} h(X_i)$$

(1)

where function h signifies the evaluation of a simulation result to determine if the netlist with a parameter set X_i fails or passes the test with specified criteria. If the assigned parameter set fails to produce the target conditions at the testbench output, the function will return 0. However, if the sample satisfies the conditions, the output of h will be 1 and the returned values from different sets with the correct output among N samples are accumulated and assigned to N_p. As N gets larger, the yield result becomes more accurate.

The advantage of assigning yield as an objective function is its ease of implementation. Additionally, it is also a more realistic measure of cell performance due to changing cell parameters. The accuracy of the yield analysis depends on the number of MC simulations with more simulations providing more accurate results. This comes at the cost of large MC simulation time, and therefore, a tradeoff exists between the accuracy and runtime.

To reduce long MC simulation times, Shahsavani and Pedram (2019) introduced a novel way of estimating the parametric yield without simulating the cells [21]. In their solution, the behavior of logic cells are modelled with a machine learning based approach for predicting whether a cell operates properly or not by looking at the values of a given parameter set on each MC sample. The training and test data sets for the model is obtained from MC simulations of desired logic cells and the authors achieved an average of 96% accuracy for the trained model on this alternative yield analysis.

2.2.3 Multi-Criteria Objective Function

Considering the techniques on Sects. 2.2.1 and 2.2.2, it is possible to utilize both of margin and yield scores for the optimization process. Each of the scores can be added up after multiplying by an individual weight as shown in (2).

$$f_c(X) = w_m * M(X) + w_y * Y(X)$$

(2)

In the score function f_c, critical margin score, margin weight, parametric yield, yield weight and a set of cell parameter values are represented as M, w_m, Y, w_y and X. $M(X)$ and $Y(X)$ correspond to evaluated margin and yield scores of parameter values X. Additionally, these scores are normalized while $0 \leq w_m + w_y \leq 1$. These constraints are applied to map the cell scores between 0 and 1. Among the obtained scores, the best choice would be the one closest to the value 1 which corresponds to a cell having largest critical margin and best parametric yield.

In addition to margin and yield scores, [22] includes a current leakage L as another criteria for SFQ cells while comparing the performances of different objective functions [23] and [24]. The leakage value is obtained from a single analog simulation of a cell. Dividing average current passing through a resistance component to summed average currents passing through inductance components in a cell gives a rough estimation for this metric. In their work, the goal is set to minimize the leakage while having maximum margin and yield scores. The first score function called DEA is determined as shown in (3).

$$f_c(X) = \frac{w_m * M(X) + w_y * Y(X)}{w_l * L(X)} \tag{3}$$

Since there is a denominator part, the weights are stated as w_m, w_m, $w_m \geq \epsilon$ to avoid zero division where ϵ is a minimum weight and $\epsilon > 0$. The second score function is scalarizing function, given as (4), which is minimized during the optimization.

$$f_c(X) = \left(\frac{M - I_m}{G_m - I_m} \right)^p + \left(\frac{Y - I_y}{G_y - I_y} \right)^p + \left(\frac{L - I_l}{G_l - I_l} \right)^p \tag{4}$$

It also uses ideal values I and goal values G. I is set as the best possible values while G is defined by the user preferences. The parameter p is assigned as 3 for exponent parts and as cell parameters approach their ideal values, the related value for the terms in parenthesis approaches zero.

2.3 Existing Optimization Methods

In this section, we provide a detailed review of the prior art techniques used for margin and yield optimization of SFQ cells. Depending on the objective function and the number of parameters to be optimized, the quality of optimization results are different; thus, finding a suitable algorithm and objective function is crucial.

2.3.1 Inscribed Hyperspheres Method

The Inscribed Hyperspheres Method (IHM), one of the design centering methods, was first used by Herr and Feldman in [8] to optimize RSFQ circuit. The original algorithm, named Simplicial Approximation (SA), was developed by Director [25]. The optimization defines a sphere, having a radius which represents the critical margin of an optimized circuit. This algorithm approximates the feasible region boundaries of an n-parameter design space, aiming to increase the volume which is inside a defined convex hull.

The initial (nominal) parameter values of an SFQ circuit are considered as starting point for the optimization. The set of m boundary points for an operating region are calculated by sweeping each parameter value within its spread range, i.e. using binary search to find the points that satisfy the fail/pass criteria. n denotes the number of parameters to be optimized and m is generally assigned as $2n$ or $n + 1$ by following $m \geq n + 1$. Notice that each point corresponds to a one-dimensional parameter margin. Connecting these m points, the resulting planes form a convex hull where the largest sphere that can fit will be inscribed. The computation involves an iterative process that is achieved by an approximation where the largest tangent plane defines the direction of the next parameter value search. The calculation continues until the sphere fits into the center of a hull where the optimum parameter values are defined. After maximizing the sphere volume, the new hull is drawn for a new search to commence.

The optimized parameter values in each iteration are located at the center of the inscribed spheres, corresponding to set of parameter values with the maximum tolerance to the variations. They serve as a new design vector for the next iteration. After this iteration is completed, some circuit parameters may still exhibit larger variation than the other parameters. Therefore, additional adjustments are needed to find the optimal volume. Values of parameters with large variations are used to find a better center for the sphere. Since there are additional value adjustments on parameters, the center of a sphere may move to a different location. This process eventually produces the global maximum point for the operating region of a given circuit. As stated in their paper, standard critical margin analysis for its objective function may provide a poor choice of parameter values for a cell with large number of parameters and result in low yield. Due to high computational cost of obtaining a high-dimensional SA, this method is applicable only to problems that have a low number of design parameters.

The authors of [8] implemented IHM and simulated RSFQ circuits in MALT [3]. They performed margin optimization on 8 different cells including NOT, XOR, D Flip-Flop, T1 Flip-Flop, SFQ-DC converter, DC-SFQ converter, splitter, and confluence buffer cells. In this work, global parameters are defined as XL (which multiplies all inductors), XR (which multiplies all resistors), XI_b (which multiplies all bias currents), and XI_c (which multiplies all junction critical currents). To reduce the number of parameters, XI_c and XI_b are combined into XI_{cb}, and XR is not included due to its small effect on cell dynamics. Depending on the complexity of a cell, the authors achieved margin values of $\pm 47.25\%$ and $\pm 33.75\%$ on average with respect to XL and XI_{cb} parameters.

2.3.2 Centers-of-Gravity Method

The Centers-of-Gravity Method (CGM) is a statistical approach that seeks to compute parameter values for a target design which will maximize some figure of merit (e.,g., yield) by design centering [26]. In particular, CGM works by performing a statistical exploration over a region, which is in turn defined based on the set of design components that are utilized for the parametric yield analysis in a hyper-parameter search space. For yield calculation, a Monte Carlo analysis is utilized where the process of random sampling also contributes to determining the new position for the parameter set (in addition to estimating the quality of parameter set during the optimization). Harnisch et al applied CGM to SFQ cells and investigated the algorithm's performance using different design parameters [5]. Their algorithm starts by creating a vector P which contains parameter values of a circuit:

$$\vec{P} = p_1 \ p_2 \ \ldots \ p_k \ \ldots \ p_K \tag{5}$$

where K denotes for the number of parameters that must be optimized and represents the dimension of search space. For the nominal values in the circuit, an additional vector is formed (with a superscript "o") as shown below:

$$\vec{P^o} = p_1^o \ p_2^o \ \ldots \ p_k^o \ \ldots \ p_K^o \tag{6}$$

Next N samples are drawn around the nominal values in order to determine the parameter tolerances since the results are compared to the desired performance. These tolerance values are represented with a vector T, and the tolerance range is calculated as follows:

$$\vec{T} = t_1 \ t_2 \ \ldots \ t_k \ \ldots t_K \tag{7}$$

$$(p_j^o - t_j) \leq p_j \leq (p_j^o + t_j) \quad , \quad j = 1, \ldots, K \tag{8}$$

The position and width of the tolerance region depend on the number of parameters and the nominal vectors of parameter values within the region. The passing and failing vectors for circuit parameter values are estimated as follows:

$$\vec{G}_P = \frac{1}{N_P} \sum_{\substack{Passing \\ circuits}}^{N_P} \vec{p}_i \quad , \quad \vec{G}_F = \frac{1}{N_F} \sum_{\substack{Failing \\ circuits}}^{N_F} \vec{p}_i \tag{9}$$

where N_P and N_F denotes the numbers of passing and failing circuits (note that total number of samples N is equal to $N_P + N_F$. Since these vectors are around P^o and within the tolerance region, they also contribute to the determination of the new nominal vector. Upon acquiring $\vec{G}_P$ and $\vec{G}_F$ vectors, the new nominal vector P^{o*} can be calculated as follows:

$$\vec{P^{o*}} = \vec{P^o} + \lambda (\vec{G}_P - \vec{G}_F) \tag{10}$$

where λ, which denotes the step length, is set to $1 - Y$. Note that yield Y within the tolerance region is calculated as N_P/N. After the new vector calculation, the tolerance range is updated. Next, the new samples are obtained around the new nominal vector, and the same process is repeated until the desired results are achieved or the optimization process is terminated when there are no further improvements in yield for a selected cell. In this work, CGM was applied to optimize the SFQ buffer, splitter, D Flip-Flop, NDRO, T Flip-Flop, AND, OR, XOR, and NOT cells. Although the method targeted yield optimization, it achieved cell margin improvements as well. More precisely, it resulted in at least ±35% margin improvement on the said cells, and more importantly, on circuits such as DC-SFQ converter and an RSFQ voltage doubler.

2.3.3 Genetic Algorithms

Genetic algorithms (GA) are derived from the notion of natural selection. These algorithms simulate the selection of fittest individual within the population of candidate solutions. The offsprings are generated from the successor parents and the genes of parents are passed to the next generation. The combination of these genes create the strings to form child's chromosome which will correspond to a solution. GAs only require the evaluation of a desired fitness function, and this fundamental difference makes GAs more robust than other optimization practices. Fourie and Perold (2003) mapped the design values of SFQ cell variables into the strings, which enabled them to utilize GA as an alternative optimization method [9]. The authors used theoretical yield as a fitness function where Monte Carlo analysis is performed to evaluate the quality of solutions. Therefore, the design variable values are determined based on the solution's yield and its corresponding probability of survival. Since the yield score may create insufficient differentiation among the solutions, it is normalized to assign a value of 0 for the solution with the lowest yield and a value of 1 for the one with the highest score. All other solutions are assigned values between 0 and 1. This normalized score is then used as a fitness value. Upon determining the parent solutions and finishing the reproduction, a crossover operator is used to generate new solutions. The genetic information from the parents' strings is passed to the offspring during the crossover process. In the aforesaid work, a random process is used to produce the offsprings. Moreover, mutations are allowed to occur after the crossover. The mutation corresponds to random changes in the solution string that can create different solution varieties in the population [27]. More precisely, these random values are generated by assuming Gaussian distribution of original design values. This step helps avoid premature convergence to a not-so-good a solution.

For the RSFQ application, strings of cell components represent real values of design components such as Josephson Junctions (JJs), inductors, and resistors. After the reproduction process of these strings, random cross over points are selected among the cell components. The values of corresponding components in each individual are swapped between the individuals to create variety of value sets for an SFQ cell. For each individual, random mutations alter the component values according to normal distributions around the nominal values. After the mutation process, the said individuals will constitute the new generation of component values as candidate solutions for the remaining optimization steps.

By fine-tuning the optimization parameters in different runs, GA resulted in an average yield of 95.6% for Complementary Output Switching Logic (COSL) Set-Reset Flip-Flop. Additionally, by utilizing the same optimization parameter settings, the yield of COSL negative output OR cell was increased from 58.7% to 100%. The authors stated that the yield improvement was accompanied by corresponding critical margin improvement for all cells.

2.3.4 Particle Swarm Optimization Method

It has been shown that the Particle Swarm Optimization (PSO) method [28] can provide promising results for the optimization of superconducting logic cell families and analog circuits [7, 10]. In the algorithm, a particle represents a set of parameter values for an SFQ cell. The dimension of the search space is defined as the number of cell parameters, D, and its hypervolume is determined by the product of the difference between each parameter's UB and LB values. The search space itself is defined as the set of all possible parameter values which lie inside the said hypervolume. Initially, a number of particles are randomly placed all over the predefined search space. The algorithm tries to find the optimum point of the defined objective function inside the search space. Since the nature of the algorithm is stochastic search and starting point of the search is randomized, finding the global maximum point is not guaranteed. Thus, the particles may converge into the local maxima during the optimization.

The optimization process starts with the randomly placed particles where the location of each particle is denoted as X_i^t. The parameters t and i represent the iteration number and particle number where i is within the range between 1 and a predefined population size I, i.e., the number of particles. Variable d represents the current dimension, and it varies from 1 to D which is the total number of dimensions, i.e., the total number of parameter values, as shown in (11).

$$X_i^t = X_{i,1}^t, \ X_{i,2}^t, \ \dots, X_{i,d}^t, \ \dots, X_{i,D}^t \tag{11}$$

After assigning particle locations, each set of parameter values are evaluated according to the target objective function. Each particle keeps the track of its best location called the particle best (i.e., $pbest_i^t$). Next, the particle with the best score is selected and this current best location is represented as $gbest_j^t$. Once the best particle is chosen, the other particles move towards the best point.

In each iteration, each particle has a unique velocity towards the $gbest_j^t$ location. The velocity vector in each iteration, denoted as V_i^{t+1}, is a summation of three components, namely inertia, cognition, and social components [7] that are defined as follows:

$$V_i^{t+1} = C_0 \times V_i^t$$
$$+ C_1 \times rand(0, 1) \times \left(pbest_i^t - X_i^t\right) \qquad (12)$$
$$+ C_2 \times rand(0, 1) \times \left(gbest_j^t - X_i^t\right)$$

Inertia vector component acts as a memory where the particle remembers previous iteration's velocity and its distance from the $pbest_j^t$ location. Cognition vector component represents the weighted difference between the current location of a particle and the best location obtained by that particle. The social vector component is computed by the weighted difference between the current location and the best known location of all the particles in the swarm. These components are used to regulate the velocity aiming to achieve a balance between exploration and exploitation [15]. Once the velocity vector for all particles is obtained, the next location of each particle is calculated as shown in (13) [15].

$$X_i^{t+1} = X_i^t + V_i^{t+1} \qquad (13)$$

To evaluate the performance of the PSO method, the authors of [7] compared SFQ D Flip-Flop (DFF) and T Flip-Flop (TFF) cell results from the PSO tool with the results obtained by the SCOPE tool. It was reported that the critical margin results of the DFF from the PSO tool are -24% and $+33\%$ for lower and upper bounds while the SCOPE tool provided -30% and $+44\%$ for the bounds. For the TFF cell, the performance stated as -23% and $+23\%$ for PSO, and -18% and $+24\%$ for SCOPE. They also included improved delay and jitter information within their optimization criteria to have better cell quality.

3 Problem Statement and Proposed Solution

In this section, details of the objective function used for the margin optimization of SFQ cells are presented. Next, the exact problem statement is given, and our solution technique is presented. Simulation results that support the efficacy of the proposed solution are provided.

3.1 Circuit Scoring Method

Since the margin range does not capture the effect of asymmetric upper and lower margin values (ignoring the plus or minus sign of the margins), we propose a new objective function which accounts for both the margin range and the product of the upper and lower margins, aimed at increasing both the margin values and centering parameter values to generate a symmetrical critical margin. For instance, a first cell with lower and upper margins of -17% and $+1\%$, respectively, is scored equally to a second cell with a $\pm 9\%$ margin on both lower and upper bounds. Indeed, both of these cells will be assigned a critical margin range of 18% whereas the second cell is expected to have a higher parametric yield. The new score function for a cell f_c, which we call *centering-favored critical margin range*, is calculated using (14) where UB and LB are normalized to assume values between 0 and $+50\%$. UB and LB of a cell is determined by the combination of each parameter's UB and LB values. The common range of all parameters defines the metric value for a cell. Recall that an UB of $+50\%$ indicates that the parameter values can be increased by 50% compared to their nominal value without affecting the correct cell functionality. Similarly, a LB of 50% indicates that the parameter values can be decreased by 50% compared to their nominal value. X corresponds to a set of parameter values, i.e., location of a particle. In order to create a normalized score mapped between 0 and 100, we used 0.4 and 0.8 as the weights of the two components in the score function.

$$f_c(X) = 0.4 \times \sqrt{UB \times LB} + 0.8 \times (UB + LB) \tag{14}$$

Notice that $f_c(X)$ and standard critical margin range value yield equal values when $UB = LB$. In the example of the previous paragraph, the $f_c(X)$ value for the first cell is approximately 16.05% whereas for the second cell it is 18%, which does indicate that the second cell will have a higher parametric yield. Figure 2 shows the value of different components in (14) as a function of the set of possible upper and lower bound values. The square root shown in Fig. 2a indicates that the result will be higher if UB and LB values are close to each other. In Fig. 2b, UB and LB values are added together to define standard margin range. By using (14), the new margin score is obtained as shown in Fig. 2c. The new objective function gives smaller score to the margin ranges that have big difference between UB and LB values. For example, if a cell C_1 has 1% UB and 36% LB, its score with new objective function will be 32% while its standard margin score will be 37%. The cell C_1 will be treated almost equally to a cell C_2 where its both UB and LB values are 16%. The difference between the standard margin and the new margin values shows that if the cell margin tends to expand towards one side of margin bounds, the cell score will be smaller than the original case, which helps maintain both sides of bounds relatively identical. In this work, we refer to a set of parameter values as a particle. Once the margins for each parameter in a cell are calculated, the score of each particle, is computed using (14). Using the centering-favored objective function, an optimization algorithm targets generating a more symmetrical set of margins, hence, indirectly increasing the parametric yield as well as individual parameter margins.

3.2   Statement of the Optimization Problem

Each component in these designs has a tolerance to parameter spread (which arises due to imprecision of and imperfections during the fabrication process). When

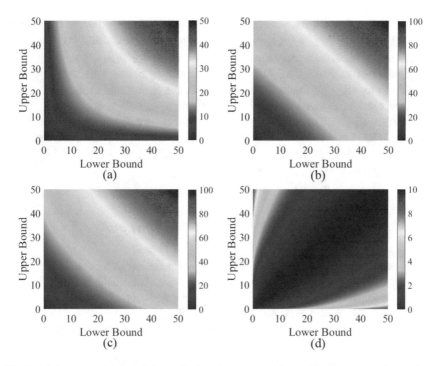

Fig. 2 (**a**) Square root value of the product of the upper and lower bounds in percentage points. This value carries the information of how well-centered the margin is and having relatively close lower and upper bounds results in a higher score. (**b**) Standard margin score. The summation of upper and lower bounds gives the quality of a margin without any deduction on the margin score. (**c**) New margin score. The combination of both square root value of the product and standard margin score allows us to prioritize the selection of parameter set with centered margins. (**d**) The difference between the new score and standard score. Scoring method penalizes the margins that are not-centered and the deduction gets higher on one-sided margins

these components are integrated on a chip, the tolerances tend to become smaller because of cell interactions. When the parameter spread is more than a component can tolerate, the correct functionality of a cell cannot be maintained. To maintain the correct functionality of cells, each component's operating margin has to be increased. Therefore, the optimization of SFQ cells is of high importance.

Margin Optimization Problem Given is an RSFQ circuit comprising many logic cells, each logic cell including a number of cell components such as Josephson Junctions (JJs), inductors, and bias resistors, and current sources. Let D denote the total number of cell components in the circuit. For each type of cell component, i.e., inductors L, resistors R, bias current I_{bias}, and critical current of JJs I_{crit}, a parameter spread in the form of a probability distribution function is defined over a desired range. Considering the inductor component type, we are given L_{min} and L_{max} where the nominal value of L (which is the mean value of the assumed normal distribution) is denoted by μ_L and calculated as $\frac{(L_{max}+L_{min})}{2}$. The corresponding standard deviation is denoted by σ_L and calculated as $\frac{(L_{max}-L_{min})}{6}$. Without loss of generality, all L component values in the circuit are selected randomly and independent of each other according to the said distribution (although it is also possible to partition all inductor components in the circuit into groups and assign the same random L value to all components in the same group so as to capture spatial correlations). The same process is used for other component types R, I_{bias}, and I_{crit}. The goal is to optimize the nominal values of circuit parameters such that they are robust to the said variations (centered properly in the feasible region) while maximizing the circuit score as defined in Sect. 3.1.

3.3   Proposed Solution: The ANPS-FW Algorithm

In this section, we describe a particle swarm optimization technique (ANPS-FW) utilizing two stochastic search algorithms, i.e., the Automatic Niching Particle Swarm Optimization (ANPSO) and the Fireworks Algorithm (FWA). We incorporate the Roulette Wheel Selection (RWS) algorithm for particle selections, which is a genetic operator used in genetic algorithms for selecting potentially useful solutions for recombination, into ANPS-FW to improve its performance. Moreover, we integrate the damping wall (DW) technique into ANPS-FW, which allows particles to approach (but not go outside) the search space by adjusting the direction and velocity of approaching particles so that the boundary regions can be searched better. Last but not least, a new objective function for optimizing SFQ logic cells which accounts for both the sum and the minimum value of lower and upper bound margins of cell parameters is proposed and adopted. The overall flow of ANPS-FW is shown in Fig. 3. The inputs to this algorithm consist of the netlist of a logic cell and the cell parameter values and spreads. The output of this algorithm is an optimized cell, with modified parameter values and improved critical margins. We refer to a set of cell parameter values as a particle and a set of particles as a swarm. A subswarm is a subset of an original swarm, obtained by partitioning the said swarm into multiple parts.

In the initialization step, the input netlist is parsed, simulated, and checked to see if the cell is working correctly, i.e., whether the initial set of parameter values result in a feasible solution. In step 1 of the algorithm, a particle corresponding to the set of nominal values of cell parameters along with a number ($S_M - 1$) of randomly generated particles are created. At the second step, each particle is assigned as a firework location. Each firework then creates S_N number of sparks and the location of each spark is recorded. In the fourth step, the brightest spark among S_N sparks is selected for each particle. In step five, a predefined number of best particles are assigned to subswarms (using the RWS algorithm). In step six, we compare the best scores of subswarms with a target margin score and decide whether to continue the search or terminate the algorithm. Additionally, we terminate the search if the

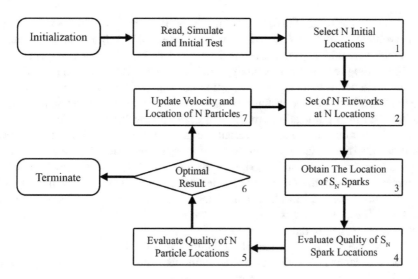

Fig. 3 Flow graph for the combination of ANPSO and FWA

total number of iterations reaches a predefined limit. If the conditions to terminate the search are not satisfied, all particles' locations are updated and the algorithm repeats starting from the second step. In the end, the algorithm reports the updated margins for individual parameters as well as the critical margin of the optimized cell. The parametric yield of the optimized cell can also be reported to verify the effectiveness of the algorithm.

The given methodology (ANPS-FW) essentially combines the Automatic Niching Particle Swarm Optimization (ANPSO) and Fireworks Algorithm (FWA) algorithms. By combining ANPSO and FWA optimization methods, a swarm approach with a partially connected topology which accommodates converging to multiple local optimum points will be established while enabling efficient global and local search. FWA will further boost the performance of ANPSO algorithm in terms of finding high-quality local search spaces. The algorithm is elaborated on in the next two subsections.

3.3.1   ANPSO in View of FWA

As stated in Sect. 2.3.4, the set of parameter values in RSFQ cell corresponds to a particle in this algorithm as well. We refer to a population of particles as a swarm, and subset of a swarm as a subswarm. ANPSO is a swarm optimization approach that moves particles within a subswarm. Every subswarm has an elite member that leads the rest of the particles to a local optimum. For each particle, the Euclidean distances of that particle to all the subswarm elites are calculated and each particle is assigned to the closest subswarm.

Similar to PSO algorithm, D represents the dimension of the search space and the hypervolume depends on UB and LB values in the ANPSO algorithm. The location of particles are initially randomized and placed to the predefined search space. Even though finding the global maximum point is not guaranteed like in PSO, ANPSO has multiple points that particles converge into. Thus, this feature might allow us to obtain better optimization results during the process. Particles are randomly placed to initialize the optimization process and each particle location is denoted as X_i^t where the parameters t and i represent the iteration number and particle number. Additionally, the variable d represents the current dimension. The corresponding equation for particle location is shown previously in (11). Every particle partially communicates with the other particles depending on their distance to the selected best locations. The best locations among the $pbest_i^t$ locations is defined as the local best, i.e., $lbest_j^t$ where j is a number which ranges from 1 to the total number of subswarms, J, within the whole PSO.

In ANPS-FW, we select local best values using the Roulette Wheel Selection (RWS) algorithm [17] rather than the traditional Rank Selection (RS) algorithm. Due to the nature of SFQ cell margin optimization, where the search space is non-convex [29], it is crucial to have an algorithm that can allow possible explorations of a large number of candidate particle locations within the search space. RS algorithm compares all of the local best points and ranks them from high to low. Afterwards, the highest score will be selected. However, in RWS, the selection can be any of the results with a non-zero probability that depends on their score. The advantage of using RWS instead of RS is that it gives a chance to the particles with a low score since they might eventually lead to a high-quality solution. An example is shown in Fig. 4a. In this figure, particle 1 has a lower score than particle 2, but it is closer to the target point, i.e., the global maximum. Therefore, it is expected that if particle 1 is selected, the global optimum can be found. However, RS algorithm chooses particle 2 over 1. On the other hand, the RWS algorithm may choose particle 1 over 2. In this case, particle 1 might have a better critical margin.

The RWS algorithm is a genetic selection algorithm that stochastically selects parents from one generation to create the basis of the next generation. The parents are assigned fitness levels and the fittest individuals have a greater chance of survival than weaker ones [17]. In the context of the ANPSO algorithm, each particle's probability value, $p(X_i^t)$, is put into a portion of a roulette wheel where size of the portion represents the particle probability. The probability of selecting each particle as $lbest_j^t$ is defined using (15). The details of centering-favored objective function are given under Sect. 2.2.1.

$$p(X_i^t) = \frac{f_c(X_i^t)}{\sum_{i=1}^{I} f_c(X_i^t)} \tag{15}$$

Using Eq. (15), the probability of particles 1 and 2 are calculated as 14% and 46%, respectively, shown in Fig. 4b.

A fixed point on the roulette wheel will be randomly chosen as a particle. A single virtual spin on the wheel will complete a single particle selection, i.e., choosing

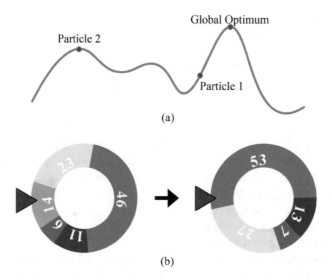

Fig. 4 (**a**) A 2D search space with two particles. Particle 2 represents a local maximum in the given search space. Even though it has higher score than particle 1, global optimum can be achieved by utilizing the parameter set of particle 1. (**b**) Roulette wheel selection. All particle probabilities are distributed on the wheel and a random particle is selected depending on its probability. Upon the selection, the remaining probability values are normalized and a new selection will occur if more than one particle is desired

one local best point. After each selection of $lbest_j^t$, the probability of choosing the remaining particles is normalized and recalculated while excluding the selected particle. It is possible that the remaining particles might have zero score on the objective function, $f_c(X_i^t)$, which will create an illegal zero division operation which represents a parameter set with zero margins. To avoid this, when the sum of these particles' score is zero, a particle is selected randomly.

In each iteration, a number of local best points are chosen. Remaining particles will then move towards the closest $lbest_j^t$ point and each of the local best particles creates a partially connected topology with the attracted particles, called a subswarm. In the ANPSO with RS selection algorithm, the local best points $lbest_j^t$ tend to stay where they are until a particle with a better score is found in the search space. However, using RWS algorithm, local best points may be selected stochastically, which allows a particle with the highest score to continue moving towards optimums. Additionally, they still record the best locations in each iterations, which allows them to keep track of local optimums. As a result, movement of the $lbest_j^t$ will additionally enable a local search near local optimums.

Once a number of local best particles is chosen, the other particles move towards the local best point inside their subswarm. In each iteration, each particle has a unique velocity towards the $lbest_j^t$ location rather than $pbest_j^t$ location given in (12) as in PSO algorithm. Even though the velocity vector in each iteration, denoted as V_i^{t+1}, is a summation of three components (inertia, cognition, and social components) [15], the inertia component utilized in this ANPSO algorithm has an adaptive feature. The related equation is defined as follows.

$$
\begin{aligned}
V_i^{t+1} = \ &\chi \times (C_0^t \times V_i^t \\
&+ C_1 \times rand(0, 1) \times \left(pbest_i^t - X_i^t\right) \\
&+ C_2 \times rand(0, 1) \times \left(lbest_j^t - X_i^t\right)
\end{aligned}
\tag{16}
$$

Inertia vector component in ANPSO algorithm acts as a memory as well and it allows the particle to remember its previous iteration's velocity. However, unlike PSO algorithm, the particle distance depends on $lbest_j^t$ location. To adjust the contribution of previous velocity, Sugeno inertia weight C_0^t, one of the inertia weight laws in Adaptive PSO, is included in the velocity calculation [30]. The equation for calculating this parameter is shown in (17). t, T, and s represent current iteration, maximum iteration, and a fixed constant, respectively. C_{0min} and C_{0max} define the limits for the inertia weight, while the fixed constant (s) adjusts the time-variant ratio for the step size of inertia weight. This approach will enable to linearly slow down particles while converging to $lbest_j^t$ which allows smaller adjustments to the location of each particle.

$$
C_0^t = C_{0min} + (C_{0max} - C_{0min}) \times \frac{1 - \frac{t}{T}}{1 + s \times \frac{t}{T}} \quad (-1 < s)
\tag{17}
$$

The contributions of cognition and social vector components are the same as in PSO algorithm but the social vector component utilizes the weighted difference between the current location and the best known location of all the particles in the subswarm. Additionally, Sugeno inertia weight C_0, cognition weight C_1, and social weight C_2 coefficients represent how much the particle's velocity is affected by the previous movement, the $pbest_i^t$ location, and the $lbest_j^t$ location, respectively. An example is shown in Fig. 5a, where the inertia, cognition, social velocity vector components as well as their summation (which is the updated next velocity vector) are depicted. Furthermore, the velocity calculation equation is modified by using the constriction factor given at (18) while creating a relation between the cognition and social weights. For high C_1 and low C_2 values, the particle will act as a self-observer, and for the low C_1 and high C_2 values, the particle will become a social-observer.

$$
\chi = \frac{2}{|2 - \phi - \sqrt{\phi^2 - 4 \times \phi}|} \quad (4 < C_1 + C_2 = \phi)
\tag{18}
$$

Upon obtaining the velocity vector for all particles, the next location of each particle is calculated as previously shown in (13) [15].

If the next location X_i^{t+1} for a particle does not lie within the search space, the location will be mapped to the search space. Four different location correction algorithms are considered in this paper: reflecting wall, damping wall, absorbing wall and invisible wall [19]. Reflecting wall changes the velocity vector sign in

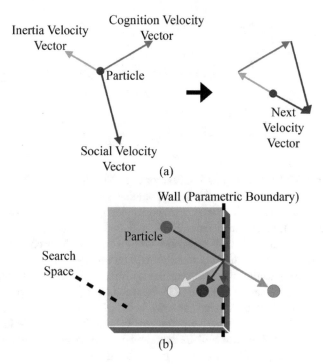

Fig. 5 (**a**) Particle's next velocity vector with its weighted components and next location. (**b**) Next velocity vectors with reflecting wall (yellow), damping wall (red), absorbing wall (blue), and invisible wall (green) techniques

the opposite direction when the particle leaves one of the dimensions of the search space. This technique might lead to a non-convergent process where the particle keeps bouncing from the boundary. Damping wall can be considered as a subset of reflecting wall where the particle bounces from the boundary but with a decreased velocity with a random coefficient between 0 and 1. Moreover, absorbing wall removes the excess velocity and keeps the particle on the boundary. As a result, the particle might skip the local search around the boundary region. Invisible wall allows the particle to go through the wall for one iteration. The illustration of these techniques are shown in Fig. 5b. By following the previously mentioned statements, DW technique is selected among the other boundary correction techniques for the SFQ cells due to two reasons. The first reason is that DW keeps the particle within the search space, unlike invisible wall where we do not want their cell parameter to surpass the limits of search space. Additionally, DW maintains the particle's direction towards the boundary point while reducing its excess velocity and enabling better search of boundary regions [19]. To compute the updated location, an initial DW time-variant constant ($\lambda_{i,d}$) is declared as (19) where d represents each parameter number. This constant modifies the excessive velocity for the next iteration as shown in (20). As a final step, all of the parameters' result

$X'_{i,d}$ will be assigned back into X_i^{t+1}.

$$\lambda_{i,d} = \frac{XUB_{i,d} - XLB_{i,d}}{max(|XUB_{i,d}|, |XLB_{i,d}|)} \tag{19}$$

$$X'_{i,d} = \begin{cases} XUB_{i,d} - \lambda_{i,d} \times (X_{i,d}^t - XUB_{i,d}), & \text{if } X_{i,d}^t > XUB_{i,d} \\ XLB_{i,d} + \lambda_{i,d} \times (XLB_{i,d} - X_{i,d}^t), & \text{if } X_{i,d}^t < XLB_{i,d} \\ X_{i,d}^t, & \text{otherwise} \end{cases} \tag{20}$$

After the calculation of a new location for each particle, the margin calculation is performed and the score of a cell netlist is obtained. Before comparing the result of the objective function with $pbest_i^t$, Fireworks Algorithm (FWA) is used for local search. The location obtained from FWA will be compared to particle's current $pbest_i^t$. At this point, the RWS algorithm will be used to determine $lbest_j^t$ among the whole population and the next iteration will be initiated.

In their proposed algorithm, during the optimization process at each iteration t, the average Euclidean distance, $d(X_i^t)$, between $lbest_i^t$ and each particle's X_i^t is expected to be bigger than Δ value as shown in (21). The absolute value calculations in (21) are used to handle negative parameter values. If any particle has a shorter distance than Δ, its location will be randomized and assigned within the search space, but the particle will still keep its $pbest_i^t$ location.

$$d(X_i^t) = mean\left(\frac{|lbest_j^t - X_{i,d}^t|}{|XUB_{i,d} - XLB_{i,d}|}\right) > \Delta \tag{21}$$

When calculating the particle locations, it can be aimed to generate identical parameter values for symmetrical cell structures, such as SQUIDs. In this case, such parameters will be grouped together, and the changes will be applied to all the parameters in a group when calculating the particle locations. Consequently, a correction function will be called and the grouped parameters will be assigned same values. As a result of parameter grouping, the dimension of search space is reduced, boosting the optimization process.

3.3.2 FWA in View of ANSPO

As mentioned in Sect. 3.3.1, during the ANPSO process, we use the FWA algorithm for performing local search before the evaluation of particles. Each particle generated using the ANPSO algorithm will be considered as a firework. The location of fireworks will also be obtained from ANPSO in the optimization process rather than using the probabilistic approach given at [15]. Initially, the best (Y_{max}^t) and the worst (Y_{min}^t) fireworks are selected using (22).

$$Y_{max}^t = max(f_c(X_i^t))$$
$$Y_{min}^t = min(f_c(X_i^t))$$

(22)

After finding the best and the worst fireworks, the number of sparks (S_i^t) will be calculated using (23). The number of sparks is proportional to the quality of a firework and each spark represents a different parameter set for a cell. Parameter ξ is defined as a small constant to prevent zero division. m is a positive integer called the control number.

$$S_i^t = m \times \frac{f_c(X_i^t) - Y_{min}^t + \xi}{(\sum_{i=1}^{I} f_c(X_i^t) - Y_{min}^t) + \xi}$$

(23)

Spark limitations for each S_i^t are defined as in (24). Parameters a and b are predefined for lower and upper bounds, respectively.

$$\hat{S}_i^t = \begin{cases} round(a \times m), & \text{if } S_i^t < a \times m \\ round(b \times m), & \text{if } S_i^t < b \times m \quad (0 < a < b < 1) \\ round(S_i), & \text{otherwise} \end{cases}$$

(24)

In addition to the spark amount, the amplitude of firework explosion should be considered. For high-quality fireworks, the explosion range should be small. The size of the explosion carries information about population convergence as it corresponds to the quality of current fireworks. Every parameter of the SFQ cell is considered to be normally distributed, and standard deviation value (σ) is a predefined percentage (p) of parameter's nominal value. The control number ($\hat{A}$) for the amplitude of firework explosion (A_i^t) is calculated as $(1 - p) \cdot X_i^t$. The control number enables to find the extreme values which cannot be achieved in the normal distribution and it is used as shown in (25).

$$A_i^t = \hat{A} \times \frac{Y_{max}^t - f_c(X_i^t) + \xi}{(\sum_{i=1}^{I} Y_{max}^t - f_c(X_i^t)) + \xi}$$

(25)

To generate diverse sparks (different parameter sets) in the explosion, the parameters are chosen randomly. For each spark, different vector z_n^t will be defined as shown in (26). The vector size is equal to the number of cell parameters which are to be optimized. By using the spark diversity parameter, the displacement for the explosion effect is calculated by using (27). The vector z_n^t is created for every spark and the parameter n represents the current spark while it varies from 1 to $\hat{S}_i^t$.

$$z_n^t = \begin{bmatrix} round(rand(0, 1)) & \rightarrow 1 \\ round(rand(0, 1)) & \rightarrow 2 \\ \vdots & \vdots \\ round(rand(0, 1)) & \rightarrow D \end{bmatrix}$$

(26)

$$h_n^t = z_n^t \times A_i^t \times rand(-1, 1) \tag{27}$$

Firework location will be imported for each spark location before the explosion. Afterwards, a unique displacement value is added to the initial location of each sparks as shown in (28). The new location is mapped back to the search space by using previously mentioned DW technique in the Sect. 3.3.1. Additionally, a correction function will be called upon the grouped parameters after every location assignment. As a result, the symmetrical structures will be maintained.

$$\bar{x}_n^t = X_i^t$$
$$\bar{x}_n' = \bar{x}_n^t + h_n^t \tag{28}$$

The objective function will be called for $\bar{x}_n'$ and each spark quality (for all n values) will be determined in the same way. In the final step, the brightest spark will be chosen and it will be assigned back to the location of current particle (X_i^t) on ANPSO algorithm. As shown in Fig. 3, after multiple iterations of the ANPSO and FWA algorithms, an optimized set of parameter values is obtained with maximized score function, i.e, an optimized cell with enhanced and well-centered critical margins is generated.

4 Results of ANPS-FW

To evaluate the efficacy of ANPS-FW algorithm, we used the cell library (v1.5) [31] as a baseline solution, and applied both plain PSO [7] and the hybrid optimization algorithm (ANPS-FW) to 6 cells from this manually optimized library. Fan-in and fan-out cells are selected to be Josephson Transmission Line (JTL) cells [1]. The terminal nodes of the JTL cells are grounded through a load resistor. We implemented the hybrid algorithm in MATLAB and used JSIM (Josephson SIMulator) [32] for simulating cells.

In the experiments, C_1 and C_2 values are kept the same for both plain PSO and ANPS-FW. Step size (ss) used for finding the margin range is set as 1% of the nominal value of each parameter. Iteration limit for each optimization is set to 50 for ANPS-FW. To conduct a fair experiment, we increased the iteration limits for the plain PSO to match the run time. The rest of the parameters of their optimization algorithm are listed in Table 1. The population of particles is set to 20 and the number of subswarms that divide the population into smaller groups is assigned as 4. Moreover, the minimum and maximum spark numbers are picked as 1 ($m * a$) and 5 ($m * b$). The reason for picking minimum spark number as 1 is to have at least one sample around a low quality firework. With these numbers, we aim to achieve a balanced exploration and exploitation for finding the solutions. To ensure convergence during exploration, we assign recommended values to the cognition and social weights (C_1 and C_2), which correspond to acceleration coefficients in the

Table 1 Parameter values
used in the optimization
process of ANPS-FW

Parameter	Value	Parameter	Value
I	20	m	10
J	4	a	0.1
$C_{0_{min}}$	0.4	b	0.5
$C_{0_{max}}$	0.9	T	50
C_1	2.05	s	1
C_2	2.05	ss	1%

optimization process [33]. The effect of adaptive inertia to the velocity of particles is limited to be between 0.4 and 0.9, which is decreased during the optimization process to help exploitation in the late iterations. In this work, the target margin is set to be 100%. The standard deviation, σ, is defined as 20% for all parameter types during the optimization process, and every parameter is taken into an account for the optimization of critical margin. ξ and Δ values are assigned as 0.0001 and 0.5%, respectively. We have also defined a velocity limitation to be 20% of each parameters' search range. It allows to limit the amount of a parameter's maximum value adjustment in each iteration and decreases the convergence speed to achieve a more effective global search.

For evaluating the parametric yield of each cell, we have performed 50,000 Monte Carlo (MC) simulations for each of the baseline and optimized cells [2]. During the MC simulations, the standard deviations, corresponding to Josephson Junctions (JJ) σ_{JJ}, inductances σ_L, and resistances σ_R are set as 12%, 13%, and 14%, respectively. The input patterns which are applied in the optimization process and MC analysis for each cell cover every possible combination. SFQ pulses that appear on the output nodes interpreted as logic-1 while the pulse absence is considered as logic-0. Since the baseline cell library is optimized to work at 50 GHz, all the clock-Q cell delays should be less than 20 picosecond. Based on this observation, we monitor a time window of 40 picoseconds at the output to determine whether a cell functions correctly or not [31]. It means that if an output pulse does not appear within the desired time interval, the cell functionality will be considered incorrect. SFQ pulses are observed on the node of intermediate JTL cell connection at the output.

The simulation results, including the baseline and improved critical margins and parametric yield values are reported in Table 2. The negative and positive margin values report the lower and upper bound critical margins for each cell. As observed, the proposed method consistently improves the upper and lower bounds of critical margins for all the cells. Improvements in critical margin values are further verified by observing improvements in the parametric yield of the cells. As a result of the optimization process, the parameter values obtained for each cell are more robust to variations in the fabrication process, i.e., have a larger critical margin. Hence, during Monte Carlo simulations, the number of cells with correct behavior is improved, as much as 24% for some cells. Furthermore, the proposed optimization algorithm achieves more symmetrical critical margin values for most of the cells since the

Table 2 Results of critical margin and parametric yield for baseline [31] and optimized cells

Results						
	Baseline [31]		Plain PSO [7]		Proposed	
Cell	Margin (%)	Yield (%)	Margin (%)	Yield (%)	Margin (%)	Yield (%)
OR	[−23,+23]	73.00%	[−31,+25]	76.45	[−35,+37]	93.50%
AND	[−44,+38]	97.00%	[−44,+47]	97.92	[−46,+49]	98.20%
XOR	[−14,+29]	60.60%	[−20,+41]	77.66	[−26,+41]	83.90%
NOT	[−25,+26]	89.70%	[−32,+38]	93.59	[−37,+38]	94.30%
DFF	[−20,+19]	74.20%	[−28,+29]	88.49	[−29,+32]	91.40%
NDRO	[−26,+10]	61.50%	[−26,+23]	65.93	[−32,+32]	86.00%

Improvements				
	Proposed vs baseline [31]		Proposed vs plain PSO [7]	
Cell	Margin (%)	Yield (%)	Margin (%)	Yield (%)
OR	[+12,+14]	20.50%	[+4,+12]	17.05
AND	[+2,+11]	1.20%	[+2,+2]	0.28
XOR	[+12,+12]	23.30%	[+6,+0]	6.24
NOT	[+12,+12]	4.60%	[+5,+0]	0.71
DFF	[+9,+13]	17.20%	[+1,+3]	2.91
NDRO	[+6,+22]	24.50%	[+6,+9]	20.07
Average	[+8.83,+14]	15.22%	[+4,+4.3]	7.88

nominal values of the parameters are centered within the range of possible values when it is compared to the plain PSO results. As an example, the critical margins for the NDRO cell are improved from −26% and +10%, to −32% and +32%, respectively. In the optimized NDRO cell, parameter values can now change to 68% or 132% of their nominal value, while still obtaining a higher parametric yield, when compared with the baseline cell. As shown in Table 2, the critical margin improvement for the NDRO cell is stated as +6% and +22%. In other words, the NDRO cell now has +6% more margin for its lower bound and +22% more margin for its upper bound. The average improvement for all cells is observed as +8.83% and +14%. The proposed algorithm has better improvements for critical margins of the cells when compared with the plain PSO except for NOT and XOR cells on their upper bounds. The plain PSO approach primarily focuses on improving the margin range while not directly considering the fact that whether the resultant margins are centered or one-sided. Also, due to its greedy nature stemming from RS algorithm, plain PSO cannot avoid premature convergence to a local optimum. The proposed algorithm further improves all critical margins 8.3% on average.

The primary reason behind the superior performance of ANPS-FW compared with the plain PSO approach is the independent search achieved by different subswarms of ANPSO and an additional local search on each particle location done by FW algorithm. During the optimization process, their critical scoring method also helps to obtain/maintain critical margins that are symmetrical on both upper and lower bound values.

We observed that approximately within the first 20 iterations of ANPS-FW, the particles tend to converge to near-optimal solutions. The rest of the iterations mostly focuses on centering the resulting margin range with the help of FWA while still trying to improve total critical margin range. The convergence speed is time-variant since the proposed algorithm uses an adaptive inertia weight. Adaptive inertia weighting enables improving the quality of results through minor changes of the parameter values during the optimization process in the final stages.

Note that the baseline solution is an already optimized cell library, offering good initial locations for the particles, which actually helps the plain PSO significantly, as their approach is capable of searching other local maximums independently. Therefore, it is still possible to observe similar results depending on the initial critical margin range of SFQ cells. However, our approach manages to outperform the PSO algorithm for all the cells in terms of total margin ranges and parametric yield values.

Notice that because there are structural differences between cells and each cell has a different number of R, L, and JJ elements with various parameter values, the same amount of increase in the critical margin of cells improves the parametric yield values of different designs by different degrees.

5 Conclusion

To achieve balanced and improved cell parameters, the hybrid method for RSFQ cell library optimization has been demonstrated and reported. ANPSO method scans the global search space while FWA improves the chances of finding best parameter set within a local area of each particle. By giving smaller score to not-centered critical margin range, particles are pushed to equalize lower and upper bounds during the optimization process. The increased critical margin results of sample cells also show the improvement on the yield.

References

1. K.K. Likharev, V. K. Semenov, RSFQ logic/memory family: a new josephson-junction technology for sub-terahertz-clock-frequency digital systems. IEEE Trans. Appl. Supercond. 1(1), 3–28 (1991)
2. C.A. Hamilton, K.C. Gilbert, Margins and yield in single flux quantum logic. IEEE Trans, Appl. Supercond. 1(4), 157–163 (1991)
3. S. Whiteley, Josephson junctions in spice3. IEEE Trans. Magnet. 27(2), 2902–2905 (1991)
4. S. Polonsky, P. Shevchenko, A. Kirichenko, D. Zinoviev, A. Rylyakov, PSCAN'96: new software for simulation and optimization of complex RSFQ circuits. IEEE Trans. Appl. Supercond. 7(2), 2685–2689 (1997)
5. T. Harnisch, J. Kunert, H. Toepfer, H. Uhlmann, Design centering methods for yield optimization of cryoelectronic circuits. IEEE Trans. Appl. Supercond. 7(2), 3434–3437 (1997)

6. N. Mori, A. Akahori, T. Sato, N. Takeuchi, A. Fujimaki, H. Hayakawa, A new optimization procedure for single flux quantum circuits. Phys. C **357–360**, 1557–1560 (2001). https://www.sciencedirect.com/science/article/pii/S0921453401005470

7. Y. Tukel, A. Bozbey, C.A. Tunc, Development of an optimization tool for RSFQ digital cell library using particle swarm. IEEE Trans. Appl. Supercond. **23**(3), 1 700 805–1 700 805 (2013)

8. Q. Kerr, M. Feldman, Multiparameter optimization of rsfq circuits using the method of inscribed hyperspheres. IEEE Trans. Appl. Supercond. **5**(2), 3337–3340 (1995)

9. C. Fourie, W. Perold, Comparison of genetic algorithms to other optimization techniques for raising circuit yield in superconducting digital circuits. IEEE Trans. Appl. Supercond. **13**(2), 511–514 (2003)

10. M.A. Karamuftuoglu, S. Demirhan, Y. Komura, M.E. Çelik, M. Tanaka, A. Bozbey, A. Fujimaki, Development of an optimizer for vortex transitional memory using particle swarm optimization. IEEE Trans. Appl. Supercond. **26**(8), 1–6 (2016)

11. P. Fourie, A. Groenwold, The particle swarm optimization algorithm in size and shape optimization. Struct. Multidisc. Optim. **23**, 259–267 (2002)

12. T. Hendtlass, *The Particle Swarm Algorithm* (Springer, Berlin, 2008), pp. 1029–1062. https://doi.org/10.1007/978-3-540-78293-3_23

13. Y. Tan, Y. Shi, H. Mo, *Proceedings of the Advances in Swarm Intelligence: 4th International Conference*, ICSI, Harbin, June 12–15, Part I, vol. 7928 (2013). http://link.springer.com/10.1007/978-3-642-38703-6

14. R. Brits, A. Engelbrecht, F.V. den Bergh, A niching particle swarm optimizer, in *Proceedings of the 4th Asia-Pacific Conference on Simulated Evolution* (2002)

15. Y. Tan, *Fireworks Algorithm for Optimization in Advances in Swarm Intelligence*, ed. by Y. Tan, Y. Shi, K.C. Tan (Springer, Berlin, 2010). https://doi.org/10.1007%2F978-3-642-13495-1

16. A. Nickabadi, M.M. Ebadzadeh, R. Safabakhsh, A novel particle swarm optimization algorithm with adaptive inertia weight. Appl. Soft Comput. **11**, 3658–3670 (2011)

17. J.H. Holland, *Adaptation in Natural and Artificial Systems: An Introductory Analysis with Applications to Biology, Control, and Artificial Intelligence* (University of Michigan Press, Ann Arbor, 1975). https://books.google.com/books?id=JE5RAAAAMAAJ

18. V. Perlibakas, Distance measures for PCA-based face recognition. Pattern Recogn. Lett. **25**(6), 711–724 (2004). http://www.sciencedirect.com/science/article/pii/S0167865504000248

19. T. Huang, A.S. Mohan, A hybrid boundary condition for robust particle swarm optimization. IEEE Antennas Wirel. Propag. Lett. **4**, 112–117 (2005)

20. R. Kleiner, D. Koelle, F. Ludwig, J. Clarke, Superconducting quantum interference devices: state of the art and applications. Proc. IEEE **92**(10), 1534–1548 (2004)

21. S. Nazar Shahsavani, M. Pedram, A hyper-parameter based margin calculation algorithm for single flux quantum logic cells, in *2019 IEEE Computer Society Annual Symposium on VLSI (ISVLSI)* (2019), pp. 645–650

22. F.G. Ortmann, A. van der Merwe, H.R. Gerber, C.J. Fourie, A comparison of multi-criteria evaluation methods for RSFQ circuit optimization. IEEE Trans. Appl. Supercond. **21**(3), 801–804 (2011)

23. A. Charnes, W. Cooper, E. Rhodes, Measuring the efficiency of decision making units. Eur. J. Oper. Res. **2**(6), 429–444 (1978). https://www.sciencedirect.com/science/article/pii/0377221778901388

24. T.J. Stewart, A multi-criteria decision support system for R&D project selection. J. Oper. Res. Soc. **42**(1), 17–26 (1991). http://www.jstor.org/stable/2582992

25. S. Director, G. Hachtel, The simplicial approximation approach to design centering. IEEE Trans. Circuits Syst. **24**(7), 363–372 (1977)

26. R. Soin, R. Spence, Statistical exploration approach to design centring. IEE Proc. G Electron. Circuits Syst. **127**(6), 260–269 (1980), cited By 55

27. D.E. Goldberg, *Genetic Algorithms in Search, Optimization and Machine Learning*, 1st ed. (Addison-Wesley Longman Publishing Co. Inc., Boston, 1989)

28. J. Kennedy, R. Eberhart, Particle swarm optimization, in *Proceedings of the IEEE International Conference on Neural Networks* (1995), pp. 1942–1948
29. L.C. Muller, RSFQ digital circuit design automation and optimisation. Accepted: 2015-05-20T09:27:37Z. https://scholar.sun.ac.za:443/handle/10019.1/96808
30. T. Takagi, M. Sugeno, Fuzzy identification of systems and its applications to modeling and control. IEEE Trans. Syst. Man Cybernet. **SMC-15**(1), 116–132 (1985)
31. S.U.N. Magnetics, sunmagnetics/RSFQlib (2020). Original-date: 2018-05-04T09:55:29Z. https://github.com/sunmagnetics/RSFQlib
32. E.S. Fang, T. Van Duzer, A josephson integrated circuit simulator (JSIM) for superconductive electronics application (1989), pp. 407–410
33. D. Bratton, J. Kennedy, Defining a standard for particle swarm optimization, in *2007 IEEE Swarm Intelligence Symposium* (2007), pp. 120–127

Hardware Security of SFQ Circuits

Tahereh Jabbari, Yerzhan Mustafa, Eby G. Friedman, and Selçuk Köse

1 Principles of SFQ Logic

The fundamental principles of single flux quantum (SFQ) logic are described in this section. The operation of a Josephson junction (JJ) is described in Sect. 1.1. SFQ logic is explained in Sect. 1.2. The principles of hardware security of SFQ circuits are described in Sect. 1.3.

1.1 Josephson Junctions

Superconductive materials exhibit zero electrical resistance when cooled below a temperature known as the critical temperature T_C [1]. The Josephson effect is described as quantum tunneling in a superconductor across a thin insulator barrier by overlap of the wave function of a Cooper pair in two superconductive layers [2]. The operation of a JJ is based on this effect. A JJ which consists of two superconductive niobium layers separated by a thin layer of oxide [3] is the primary active device in superconductive electronics. A JJ loses superconductivity when the bias current, temperature, or magnetic field exceeds, respectively, a critical current I_C, critical temperature T_C, or critical magnetic field B_C. The structure of a JJ is illustrated in Fig. 1a. JJs are modeled as a resistively and capacitively shunted junction (RCSJ) which is a depicted in Fig. 1b. The I–V characteristics of a junction are shown in Fig. 1c.

T. Jabbari (✉) · Y. Mustafa · E. G. Friedman · S. Köse
Department of Electrical and Computer Engineering, University of Rochester, Rochester, NY, USA
e-mail: tjabbari@ur.rochester.edu; ymustafa@ur.rochester.edu; friedman@ece.rochester.edu; selcuk.kose@rochester.edu

© The Author(s), under exclusive license to Springer Nature Switzerland AG 2023   135
R. O. Topaloglu (ed.), *Design Automation of Quantum Computers*,
https://doi.org/10.1007/978-3-031-15699-1_7

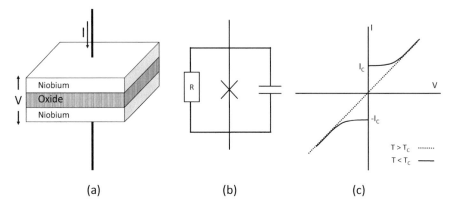

Fig. 1 Josephson junction, (**a**) structure, (**b**) RCSJ model, and (**c**) I–V relationship

1.2 SFQ Logic

An SFQ circuit consists of JJs and inductors. In SFQ circuits, information is transferred in the form of picosecond duration voltage pulses $V(t)$ within a quantized area [4–6]. Elementary logic gates in this circuit family can generate, pass, store, and reproduce picosecond voltage pulses. Switching a JJ is described as a 2π change in phase, producing a voltage pulse equal to a quantum of flux ($\phi_\circ = 2.07 \times 10^{-15}$ $V{\cdot}s$) [7],

$$\int V(t)dt = \phi_\circ \equiv \frac{h}{2e} , \qquad (1)$$

where $\phi_\circ$ is a single flux quantum, and h and e are, respectively, the Planck constant and electron charge.

The existence of a flux quantum represents a logic '1,' whereas the absence of a pulse is a '0.' In SFQ circuits, a clock signal is required for most logic gates (except for splitters, Josephson transmission lines (JTLs), buffers, and mergers [6, 8–10]). In these clockless gates, the propagation delay is the delay of the output with respect to the input. Alternatively, in clocked gates, the incoming SFQ pulse changes the internal state of an SFQ gate but does not change the output. The output changes only when a clock pulse arrives at a gate. The propagation delay is measured as the time elapsed after arrival of the clock pulse. SFQ gates are, therefore, similar to CMOS logic gates combined with an edge triggered flip flop.

1.3 Hardware Security of SFQ Circuits

VLSI complexity superconductive SFQ systems is one of the most promising beyond CMOS technologies for ultra-low power and ultra-high speed digital applications [5, 6]. Significant developments in the design and fabrication of superconductive electronics have resulted in device densities exceeding 600,000 Josephson junctions/cm^2 [11, 12]. Josephson junctions in SFQ circuits propagate an SFQ pulse through logic gates operating at switching speeds on the order of picoseconds, while dissipating power below 10^{-19} J [4, 13–18]. An SFQ-based arithmetic logic unit has been demonstrated to operate at frequencies approaching 80 GHz with an 8 bit SFQ datapath [19, 20].

Prospective exascale computing systems based on VLSI complexity SFQ circuits are expected to be used not only for high performance computing but also for critical security tasks. Hardware security for superconductive technology [21, 22] and novel techniques for providing trustworthy hardware based on SFQ circuits are therefore necessary. Security aware design methodologies for this technology are currently not well established. Recent progress in the fabrication and design of SFQ circuits strengthens the need for hardware security techniques targeting SFQ circuits. Furthermore, SFQ technology exhibits unique advantages and challenges, which should be considered when developing these hardware security techniques.

Due to the increasing complexity of modern systems-on-chip with advanced fabrication capabilities and higher manufacturing costs, many semiconductor companies have become fabless [23]. These fabless companies design the integrated circuits in-house while utilizing external foundries for fabrication, manufacturing, and integration. Although the cost of the IC production supply chain can be reduced by outsourcing certain processes to external foundries, this process also introduces security vulnerabilities into the systems integration design flow.

With an increasingly distributed IC production supply chain, different stages of the supply chain have become vulnerable to a number of attack vectors, such as counterfeiting, reverse engineering (RE), and intellectual property (IP) piracy. An attacker may insert a hardware Trojan at any point during the design, fabrication, manufacturing, or integration of an IC—either as on-chip circuitry or as an external component to perform additional malicious operations. Other vulnerabilities that can be introduced during the IC production supply chain are IC counterfeiting [21, 24], theft of IC masks [25], overproduction of ICs [26], and insertion of hardware Trojans [27].

Technology companies annually lose up to $4 billion due to IP violations in semiconductor technology [28, 29]. Hardware security has been established to mitigate the risks of piracy, counterfeiting, reverse engineering, and side-channel attacks [30]. If the functionality of an IC can be hidden while the IC passes through the different, potentially untrustworthy phases of the design flow, these attacks can potentially be thwarted. It is therefore important to an IC design company to protect this design flow. Counterfeiting is typically thwarted by IC camouflaging [22] or logic locking to prevent RE or by including a watermark to identify counterfeit ICs. Logic locking also provides protection against piracy and overproduction attacks.

Reverse engineering poses a major challenge to hardware security. RE is the process of analyzing the layout and functionality of a system to extract the gate-level netlist. RE can be performed as a non-invasive attack or as an invasive attack. Non-invasive RE attacks can be performed in combination with side-channel attacks where an attacker collects certain side-channel emanations such as power consumption [31–33], electromagnetic (EM) signals [34], or timing information to deduce the functionality of a circuit. In non-invasive RE attacks, an attacker does not leave an obvious footprint, making the attack difficult to detect. Alternatively, an invasive RE attack requires more advanced imaging and circuit analysis tools and may require several steps to extract the netlist. Invasive attacks typically can recover the netlist more accurately than non-invasive attacks. The initial step of an invasive RE attack is product teardown to identify the external characteristics of the product and package (*e.g.*, the pin arrangement). The next step –system level RE– analyzes the operations, functions, and timing characteristics of the interconnect paths. In the following step –process analysis– the structure and materials used for fabrication are examined. In the final step –circuit extraction– the gate-level schematic and netlist of the design are extracted. The cost and time necessary for RE attacks significantly increase with each step [35]. RE can be used to obtain confidential information about the design to recreate the gate-level netlist, allowing counterfeit ICs to be built, among other nefarious schemes.

IC camouflaging and logic locking, respectively, a layout technique and a circuit technique [21, 22], are widely used to mitigate the threat of RE attacks on hardware. The choice between IC camouflaging and logic locking depends upon the access of the expected attackers to the necessary resources. Both techniques, however, can be simultaneously used in an SFQ circuit. IC camouflaging in SFQ circuits obstructs the reverse engineering process by introducing dummy (i.e., redundant) JJs into a layout [22]. In camouflaged SFQ cells, normal and dummy JJs are both used. When certain JJs are necessary to maintain correct logical operation, the layout is slightly changed and dummy JJs are replaced with normal JJs. This technique relies on making these JJs indistinguishable to the attacker, who extracts an incorrect netlist. Distinguishing between a required JJ and a dummy JJ is difficult with RE attacks, which typically utilize delayering and analysis of the top view image of the layout. RE can only distinguish the dummy JJs by slicing an IC and analyzing a side view image of the layout. Slicing the die to detect dummy JJs is highly challenging in SFQ circuits due to the expected large number of JJs in large scale SFQ systems, and the small difference in the thickness of the tunneling barrier between a normal and dummy JJ [22]. Logic locking introduces modifications into a circuit to prevent piracy, counterfeiting, reverse engineering, and overproduction. Logic locking hides and locks the functionality of a circuit. A valid key is required for correct functionality. Applying an incorrect key on a locked circuit produces incorrect or seemingly random behavior. Even if an attacker obtains a physical copy of a circuit, reverse engineering the circuit layout does not allow the attacker to determine the intended behavior without the valid key.

2 Design of SFQ Camouflage Cells

IC camouflaging in SFQ thwarts RE attacks by introducing camouflaged cells and dummy JJs along with regular cells into a standard cell library. In this section, dummy JJs, camouflaged SFQ AND/OR gates, and camouflaged SFQ flip flops are reviewed. The structure of a dummy JJ to thwart RE is introduced in Sect. 2.1. The use of dummy JJs in SFQ AND/OR gates is described in Sect. 2.2. The use of a dummy DFF as a JTL is introduced in Sect. 2.3.

2.1 Dummy Josephson Junction

A dummy JJ never switches into the superconducting state and always behaves as a resistor. JJs are fabricated as a sequence of Nb–AlO_X–Nb layers where the AlO_X layer is the insulator. The critical current density of a JJ depends upon the thickness of the AlO_X tunneling barrier [36]. Changing the thickness of the insulating layer and the quality of the superconductive material affects the switching characteristics of a JJ. Two approaches to fabricate a dummy JJ are considered. These approaches increase the fabrication cost by requiring two additional mask steps.

2.1.1 Method 1: Vary Insulator Thickness of JJ

The critical current density and thickness of the insulation layer depend upon the SFQ fabrication technology and the physical design rules. The thickness of AlO_X is currently about 1 nm in a standard JJ technology [36]. By increasing the thickness of AlO_X beyond the $\sim$38 nm coherence length of the Nb layer, a dummy JJ can be fabricated that always behaves as a resistor [37]. The magnitude of the resistance depends upon the insulator thickness.

While an attacker can differentiate between a true JJ and a dummy JJ by slicing the die and imaging a side view, this strategy does not scale due to the large number of JJs in a typical SFQ circuit. Hence, slicing an IC to decipher the function of every JJ is extremely challenging. Alternatively, a top view image of a dummy JJ is identical to a standard JJ. Hence, it is extremely difficult and costly to distinguish between two JJs using image-based RE.

To tune the McCumber damping parameter β of a JJ [4, 5], most of the JJs in current fabrication processes are shunted with a resistance [5]. A cross section of a JJ with a shunt resistor is shown in Fig. 2. The structure is composed of two Nb layers, a stack of Nb-Al-Al_2O_3-Nb for the JJ, a Mo layer for the shunt resistors, and Nb_2O_5 and SiO_2 for the isolation layers. A thicker insulator film yields a dummy JJ that isolates the superconductive current. The minimum thickness of Al-Al_2O_3 in a dummy JJ is 40 nm.

Fig. 2 Cross section of a
normal and dummy JJ with a
shunt resistor between the M1
and M2 layers [36]

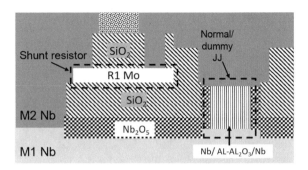

Dummy JJs are shunted with a resistor to appear identical to a normal JJ. A small shunt resistor with a dummy JJ can degrade the operation of a camouflaged cell. Decreasing the thickness of the Mo layer increases the shunt resistance and prevents deterioration of the camouflaged cell. Two approaches exist to tune the thickness of a JJ, either by addition or by elimination.

- **A double deposition process** can be used to fabricate JJs to achieve the proper critical current density for a normal JJ and to maintain the resistive behavior for a dummy JJ. The initial deposition process determines the critical current of a normal JJ. A thicker deposition layer can be used for a dummy JJ based on the coherence length. As compared to normal junctions, a dummy JJ increases the fabrication time by adding several steps. As compared to other methods to fabricate a dummy JJ, a double deposition process offers benefits that include an accurate thickness for the normal and dummy JJs and a shorter fabrication time.
- **Ion beam etching** A thick AlO_X layer is deposited for the dummy JJs. This step is followed by an ion beam etch to fabricate a normal JJ. The etching time, surface roughness, and insulator depth determine the switching characteristics of the JJ.

2.1.2 Method 2: Damage the Nb Layer

In this method, the top Nb layer of a JJ is bombarded with an ion beam to damage and degrade the properties of the Nb surface of the JJ. The ion beam smooths the surface where the damage depends upon the energy, temperature, and angle of the beam. The properties of the ions used to bombard the surface are enhanced depending upon the thickness and material of the film affected by the ion beam. Several materials produce different effects on the Nb layer. To remove an undesirable surface, an Ar or He ion beam can be used [38, 39]. Dummy JJs, fabricated using an ion beam, have a Nb layer thickness identical to a normal JJ. Furthermore, bombarding the top Nb layer with carbon ions alters the superconductive properties (e.g., eliminates the superconductive current due to the large impurity concentration within the niobium). Applying these methods, a normal JJ and dummy JJ can be separately fabricated with different and specific parameters.

A dummy JJ can therefore be included within a fabricated SFQ circuit without being recognized to thwart RE.

2.2   Camouflaged SFQ AND/OR Logic Cell

Dummy JJs are used to camouflage AND/OR SFQ gates to appear as either a two-input AND or OR gate. A camouflaged AND/OR gate exploits the structural similarity of AND and OR gates to ensure low overhead. A schematic of a camouflaged AND/OR gate with dummy JJs is shown in Fig. 3. For an AND gate, J9 and J11 are dummy JJs. Similarly, J13 is a dummy JJ within an OR gate. For analysis purposes, a dummy JJ is modeled as a large resistor in parallel with a small shunt resistor (a normal shunt resistor is 2 to 5 ohms). The high operating speed of the camouflaged cell is retained by reducing the thickness of the shunt resistor in the dummy JJ. The resistance of a shunted dummy JJ is 24 ohms. A simulation of a camouflaged AND and OR gate is shown, respectively, in Figs. 4a and 4b. A cell layout of the camouflaged AND/OR is shown in Fig. 5. The layout is based on 4.5 kA/cm^2 Hypres SFQ design rules [40].

The output delay, power, and area depend upon the number of dummy JJs. Due to the small shunt resistor in a dummy JJ, the current passing through the JJ after each input pulse is significant. Hence, the power and output delay of a camouflaged gate can be reduced by lowering the shunt resistance.

The energy dissipated by the dummy JJs in a camouflaged AND/OR is shown in Fig. 6. Averaging the energy over one clock cycle produces a power overhead of 100 pW for an AND gate with two dummy JJs, and 30 pW for an OR gate with one dummy JJ at an operating frequency of 10 GHz. The dissipated energy is approximately 2 to 5% higher than a standard SFQ OR and AND gate. The delay of the camouflaged AND/OR gates is 11 ps as compared to a delay of 10 ps for a standard AND and 7 ps for a standard OR gate. As compared to standard AND and OR gates, the area overhead of the camouflaged AND and OR gates is, respectively, approximately 15% and 10%. Since the dissipated energy is higher for camouflaged gates as compared to standard SFQ gates, one might wonder whether a side-channel

Fig. 3 Camouflaged SFQ AND-OR cell, J9 and J11 are dummy JJs for the AND gate, and J13 is a dummy JJ for the OR gate

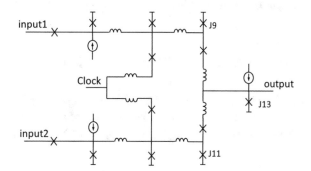

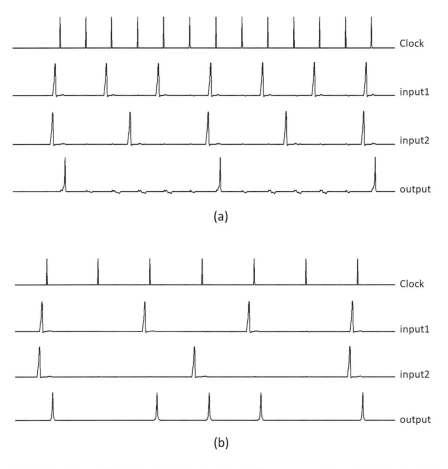

Fig. 4 Operation of SFQ cells, (**a**) AND gate (J9 and J11 are dummy JJs), and (**b**) OR gate (J13 is a dummy JJ)

attack [22] can distinguish between the two logic topologies. The energy dissipation due to JJ switching is quite low (i.e., $\sim 10^{-19}$ J), making an accurate measurement highly challenging. Hence, power side-channel attacks appear to be infeasible.

2.3 Camouflaged SFQ D Flip Flop

A Josephson transmission line (JTL) propagates a fluxon (ϕ_o) through a number of stages. A JTL improves the performance of an SFQ circuit by amplifying the SFQ pulse between logic stages. A camouflaged SFQ D flip flop (DFF) can function as

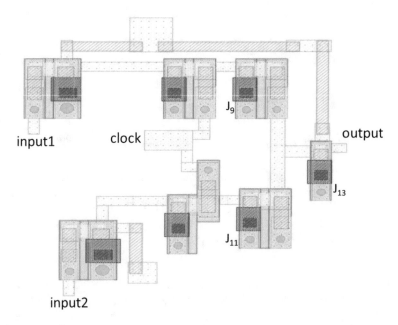

Fig. 5 Layout of a camouflaged SFQ AND-OR cell. J9 and J11 are dummy JJs for the AND gate, and J13 is a dummy JJ for the OR gate

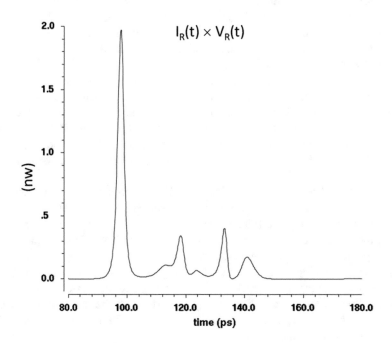

Fig. 6 Power dissipation of a dummy JJ within the AND-OR cell

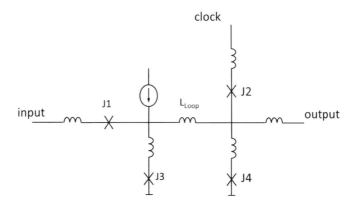

Fig. 7 Camouflaged SFQ DFF J4 is a dummy JJ

a JTL or as a standard D flip flop. A schematic of a camouflaged DFF is shown in Fig. 7. Note that the camouflaged DFF appears the same as a standard DFF.

In the camouflaged DFF shown in Fig. 7, J4 is a dummy JJ and behaves as a resistor. By adjusting the thickness of the insulating layer, the resistance is increased to lower the output delay and power dissipation. In a standard DFF, L_{Loop} is large, storing the information (i.e., bit value) when the content is read. To achieve the same physical layout, the length of the inductor in the camouflaged DFF is maintained the same as a regular DFF. Consequently, the large kinetic inductance in a camouflaged DFF produces a large output delay when functioning as a JTL. To circumvent this effect, the inductance is reduced to decrease the delay. This smaller inductance can be achieved by increasing the thickness of the kinetic inductance, thereby decreasing the overall inductance. Since a JTL is asynchronous and does not require a clock, the clock signal is eliminated by reducing the critical current of J2 by increasing the thickness of the insulator layer. Simulation results of a camouflaged DFF functioning as a JTL is shown in Fig. 8. By changing the thickness of the Nb, Mo, and insulator layers to the standard thickness, the functionality of a standard DFF can be achieved.

The output delay of a camouflaged DFF is approximately 11 ps which is roughly twice the delay of a standard JTL. This difference is attributed to the large inductance and two different input pulses—the clock and data signals. The throughput of the circuit is halved when the camouflaged DFF is part of the critical path. The current through the dummy JJ varies depending upon the clock frequency and the input signal of the camouflaged DFF. By assuming a frequency of 10 GHz for the input and clock pulses, the total power dissipated by the dummy JJ is approximately 100 pW. The energy dissipation of the camouflaged DFF is approximately the same as a standard DFF due to the different critical current of the JJs. The energy dissipated by a camouflaged DFF is approximately 2% more than

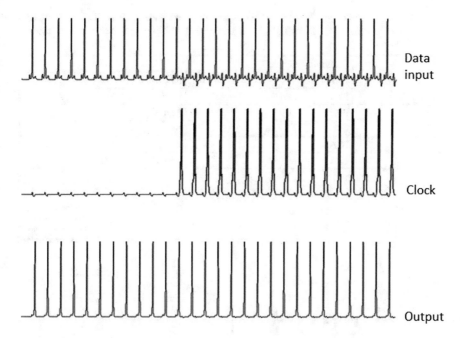

Fig. 8 Camouflaged DFF operating as a JTL. The JTL passes the input pulses regardless of the clock

a standard JTL. The top view of the camouflaged DFF is identical with a different thickness for J2 and dummy J4. A camouflaged DFF therefore exhibits the same area as a standard DFF. Furthermore, the area of a camouflaged DFF is approximately twice as large as a standard JTL.

2.4 Hardware Cost

A tradeoff between hardware security and physical area exists in camouflaged SFQ systems. ISCAS'85 benchmark circuits are used here to characterize the area and power overhead of camouflaged gates when applied to large scale SFQ circuits. Standard SFQ gates are replaced by camouflaged gates. Dummy DFFs are inserted at the inputs. The area and power overhead of the camouflaged gates as compared to standard SFQ gates is listed in Table 1. The area overhead and power overhead for these benchmarks are, respectively, approximately 40% and 30% if all of the gates are replaced with camouflaged gates, as shown in Fig. 9. The overhead is greater if additional camouflaged gates are used, providing enhanced security.

Table 1 Overhead of camouflaged SFQ cells as compared to standard SFQ cells. A camouflaged DFF behaves as a standard JTL or DFF. The camouflaged AND/OR gate behaves as a standard AND or OR cell

	Camouflaged gate					
	DFF/JTL			AND/OR		
Standard gates	Power	Delay	Area	Power	Delay	Area
DFF	0%	0%	0%	N/A		
JTL	2%	100%	100%	N/A		
AND	N/A			2 to 5%	50%	15%
OR	N/A			2 to 5%	50%	10%

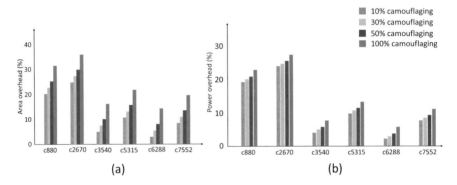

Fig. 9 Overhead of camouflaged gates, (**a**) area, and (**b**) power

3   Logic Locking

Logic locking hides the correct functionality of a circuit by introducing additional gates within the original design. In this technique, a set of key gates, key inputs, and an on-chip memory are introduced into the design to prevent attacks from the supply chain and untrusted foundries. The key gates use AND/OR gates, XOR/XNOR gates, MUX gates, and look up tables (LUT) [41]. An example of a locked circuit with key gates is shown in Fig. 10. The key inputs are K_1 and K_2 which connect to the key gates. The correct output is only produced if a correct value of the keys are applied [26]. An incorrect key used with a logic locked design causes incorrect or random operation.

Since the correct key is known only by the designer, the foundry cannot utilize any copies or overproduce and sell additional ICs without these secret keys. If the number of key values is sufficiently large, manual brute force insertion of different keys becomes infeasible. Furthermore, this technique prevents an external attacker from analyzing the structural behavior of the design even if another copy of the secured circuit is obtained.

Fig. 10 Circuit utilizing logic locking with two key gates, K_1 and K_2

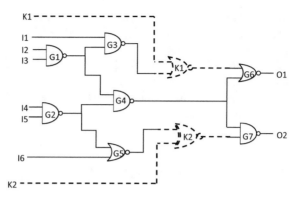

3.1 Threat Model of Attacks on Logic Locking

The primary objective of an attack on a logic locked circuit is to determine the correct value of the secret keys to decipher a functional netlist. If the keys are determined and the design is deciphered, the optimum location to insert a stealthy hardware Trojan can also be determined.

Different input patterns can be applied to both the circuit and key inputs in a brute force manner. The output of these patterns can be used to discover the correct keys. In this attack, both the locked netlist and details of the circuit design are required. The netlist can be obtained from reverse engineering a GDSII layout file, masks, or an activated functional IC. With the increasing complexity of circuits and a large number of key inputs, these attacks become highly infeasible.

The importance of hardware security in SFQ circuits is emphasized by one of the primary prospective applications of these circuits—large scale data centers typically operating with sensitive information. Countermeasures to attacks applicable to SFQ circuits are discussed in the following sections.

4 Logic Locking in SFQ Circuits

Logic locking complicates the attacks, thereby improving the security of the SFQ circuits. Existing CMOS logic locking techniques rely on introducing additional gates, look up tables, and external inputs into the design [41]. Logic locking can be similarly applied to SFQ circuits without additional modifications. The necessary gates, however –typically XOR/XNOR and multiplexers– are expensive in terms of pysical area. LUTs also require significant area. Additionally, the pinout limitations of modern superconductive ICs severely limit the size of the secret key, compromising security.

A methodology for logic locking in SFQ circuits is proposed here [21]. Rather than applying a data key, a specific current magnitude is used as the secret key.

This current is applied to specific inductances within specific gates. These locked gates exhibit incorrect operation when a different current is supplied. The internal parameters of the gates are modified and different mutual inductors are introduced, coupling the key current to the gates. The range of key currents shrinks with an increase in the number of locked cells, enhancing the security of the system. In the following section, this proposed logic locking technique is evaluated in terms of the security of SFQ circuits.

4.1 Modified OR Gate for Logic Locking

A modified OR gate is shown in Fig. 11. Mutual inductances are used to apply additional secret key current to unlock the correct functionality of an OR gate. In this circuit, the mutual inductance between L1 and L_{M1} and between L2 and L_{M2} is used to apply the key currents. The dependence of the additional secret key current on the input current and coupling coefficient K_i is

$$I_{L_i} \propto K_i I_{in}. \tag{2}$$

The inductive coupling coefficient K_n changes the current through the inductance of the internal gate. Due to the small current in the key lines, the effects of the key current on other circuit components are negligible as compared to the bias lines. To prevent any additional inductive coupling, the key lines can be placed farther from any sensitive circuit components. L1 and L2 are arbitrarily chosen as coupled inductors in the OR gate. Other gate inductors can also be used. The magnitude of

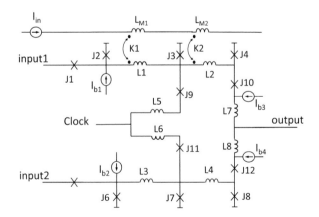

Fig. 11 Locked OR gate with mutual inductances to apply a secret key current. L1 = L3 = 15.1 pH, L2 = L4 = 3.8 pH, L5 to L8 = 5.68 pH, $L_{M1} = L_{M2} = 1$ pH, $I_{in} = 250\,\mu$A, $I_{b1,b2} = 176\,\mu$A, $I_{b3,b4} = 172\,\mu$A, $I_{C_i} = 176\,\mu$A, and $I_{c2,c6} = 250\,\mu$A (for a 10 kA/cm² technology)

the current within these inductors should be carefully chosen to maintain correct operation of the OR gate. L1 controls the current within one of the state storage loops within the OR gate. L2 affects the current within the state storage loop as well as switching junction J3. The range of the coupling coefficient between L1 and L_{M1} and between L2 and L_{M2} are, respectively, $-0.45 < K_1 < 0.45$ with zero coupling between L2 and L_{M2}, and $-0.6 < K_2 < 0.6$ with zero coupling between L1 and L_{M1}.

To unlock this OR gate, an attacker needs to determine the correct value of the key current. The correct output is only produced when the key current with a correct value is provided. Incorrect key currents coupled to inductances L_1 and L_2 produce incorrect or random circuit behavior. These incorrect key currents change the bias conditions of the SFQ storage loop by changing the current in L_1 and L_2. Incorrect operation of an SFQ OR gate is shown in Fig. 12a. The circuit incorrectly produces an output after the second and third pulse of input 1 (see Fig. 12a). The locked circuits only produce correct outputs when the appropriate magnitude of the key currents is applied to the mutual inductors with a coupling coefficient K_n. Correct operation of the circuit is shown in Fig. 12b. The correct key current is described by

$$-1 \le K_i \le 1; \qquad 0 \le I_{in} \le I_{in_{max}}, \qquad (3)$$

where $I_{in_{max}}$ is the maximum input current that can be supplied to the circuit. With a greater number of locked OR cells, the key current exponentially increases.

5   Security Characteristics of Logic Locking

An analysis of the security characteristics of the proposed technique is presented in Sect. 5.1. The area of the proposed logic locking technique based on a modified OR gate is quantified in Sect. 5.2.

5.1   Analysis of Security Characteristics

To increase the security of the proposed technique, a range of coupling coefficients for an OR gate is evaluated. By changing the coupling coefficient K_n, different fractions of the key current can be applied to the gates through the inductances. The range of K_2 within a modified OR gate for different values of K_1 is shown in Fig. 13. Each range of K_2 is a specific additional key current. The range of additional current is listed in Table 2. The key current margins are described as margins of K_2. To unlock the circuit, the correct value of K_1, range of K_2, and range of key current need to be determined. With $K_1 = 0.3$, the range of coupling coefficient K_2 is 0.25 $< K_2 < 0.35$. For smaller K_1, the circuit exhibits a large range of K_2, resulting in

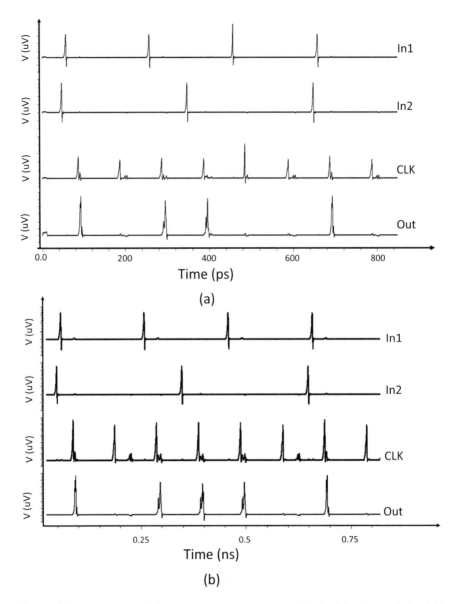

Fig. 12 OR gate operation, (**a**) incorrect current key currents with K1 = 0.5 and K2 = 0.5, and (**b**) correct current key currents with K1 = 0.3 and K2 = 0.3

lower security as compared to a higher K_1. A narrower range of K_n increases the effort required by the attacker to determine the secret key current.

Manufacturing process variations is a challenging issue in all large scale ICs. A significant tradeoff exists between circuit yield and security. To maintain proper functioning of a circuit secured by logic locking, the range of effective key currents

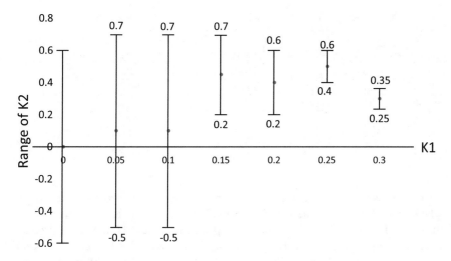

Fig. 13 Security characteristics of an OR gate for different ranges of coupling coefficients

Table 2 Range of key currents for different range of coupling coefficients

Coupling coefficient	Coupling coefficient	Current through L1	Current through L2
$K1 = 0$	$K2 = 0$	$9\,\mu A$	$-33\,\mu A$
$K1 = 0$	$-0.6 < K2 < 0.6$	$5\,\mu A$ to $13\,\mu A$	$-73\,\mu A$ to $5\,\mu A$
$K1 = 0.1$	$-0.5 < K2 < 0.7$	$12\,\mu A$ to $19\,\mu A$	$-66\,\mu A$ to $12\,\mu A$
$K1 = 0.2$	$0.2 < K2 < 0.6$	$21\,\mu A$ to $24\,\mu A$	$-17\,\mu A$ to $7\,\mu A$
$K1 = 0.3$	$0.25 < K2 < 0.35$	$27\,\mu A$ to $28\,\mu A$	$-13\,\mu A$ to $-7\,\mu A$

should be wider than any expected bias variations caused by manufacturing and the bias distribution network [42]. Process variations can improve the overall security of a logic locked system, further protecting the circuit. Unlike the intended user of an IC, the correct operation of an IC is hidden from the attacker, inhibiting a brute force attack.

Multiple locked gates can be connected to the same source of key current. These gates utilize a different magnitude and direction of inductive coupling with only a small overlap in the operational range of the key current, providing greater security. In this way, the magnitude and precision of the key currents can be increased in the case of greater manufacturing variations.

5.2 Area Overhead

An important tradeoff exists between the level of security and dedicated physical area required by the proposed logic locking technique. The area overhead of the logic locked OR gate described here is approximately 20%. ISCAS'85 benchmark

Table 3 Characteristics of ISCAS'85 benchmark circuits [43] with locked OR gates

Benchmark	# Gates	# OR gates	Area overhead with 10% locked OR gates	Area overhead with 20% locked OR gates
c880	383	90	0.5%	1%
c2670	1193	89	0.15%	0.3%
c3540	1669	160	0.2%	0.4%
c5315	2406	241	0.2%	0.4%
c6288	2406	2128	1.77%	3.6%
c7552	3512	298	0.17%	0.3%

circuits are used here to characterize the area overhead of the proposed technique when applied to large scale circuits. In the benchmark circuits listed in Table 3, the OR and NOR (OR + NOT) gates are replaced with locked OR gates to produce a narrow range for the correct key current. The number of OR gates within each benchmark circuit is listed in Table 3. Only a few locked OR gates are necessary to have a considerable impact on the security of the system. The area overhead of these benchmark circuits is also listed in Table 3, assuming 10% and 20% of the OR gates are replaced by locked OR gates. In the c6288 benchmark circuit which includes a large number of OR gates, 20% of the OR gates are replaced with locked OR gates. The area overhead is approximately 3.6%. The required area to logic lock the c6288 benchmark circuit is therefore fairly small. The area overhead is greater if additional locked gates are used to further increase the security of the circuit.

6   Attack Models on Logic Locking

One of the better known methods for attacking logic locking in CMOS circuits is a Boolean satisfiability-based attack (SAT attack) [44]. The objective of this attack method is to reduce the key space, and thus the computational time of a brute force attack. Similar attacks can be applied to SFQ circuits once a locked gate is characterized. Two attacks can target logic locking in SFQ circuits; resetting the locked OR cell and overproduction of complex locked circuits [45]. Similar to CMOS, logic locking is unable to secure SFQ circuits against these two attacks. Resetting the locked OR cell under attack is described in Sect. 6.1. The overproduction model is described in Sect. 6.2.

6.1   Attack Model #1: Increasing Correct Key Space of Locked SFQ Circuits

In the reset attack model, the range of correct key values for the locked OR cell is significantly increased by applying a specific input combination. In this section, the attack model is applied on a locked OR cell to evaluate the region of the correct key space. The threat model and attack scenario are described, respectively, in Sects. 6.1.1 and 6.1.2. Related countermeasures to prevent a reset attack are discussed in Sect. 6.1.3.

6.1.1   Threat Model

Based on Krckhoffs' principle [46], in the logic locking threat model, an attacker has access to the simulation files of the circuit, but has no knowledge of the key values. The attacker therefore knows the mutual inductances, L_1, L_2, L_{M1}, and L_{M2}, for the secret key current and can simulate the locked OR cell using arbitrary key values. The coupling coefficients are evaluated to determine the regions of operation of the locked OR cell. By sweeping the coupling coefficients from -1 to 1 at the maximum input current, the operation of the single locked OR cell (see Fig. 11) can be characterized, as shown in Fig. 14. The "C" and "I" regions represent, respectively, correct and incorrect operation. The "R" region represents

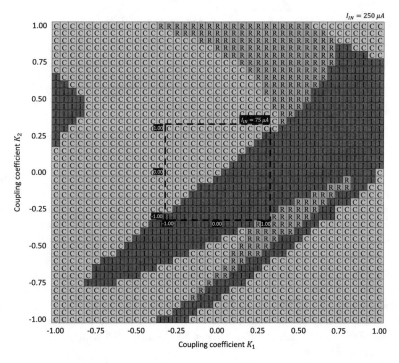

Fig. 14 Characterization of the locked OR cell when L_1 and L_2 are coupled, respectively, at input current values $I_{in} = 250\,\mu A$ and $75\,\mu A$. The "C" region corresponds to correct operation, the "I" region corresponds to incorrect operation, and the "R" region corresponds to incorrect operation which can be corrected by resetting the cell with specific inputs

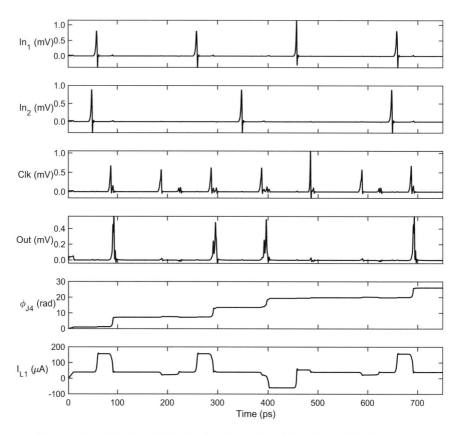

Fig. 15 Operation of the locked OR cell when $K_1 = K_2 = 0.5$ and $I_{in} = 250\,\mu A$

incorrect operation which can be transformed into correct operation as a result of the proposed attack.

For certain coupling coefficients, the locked OR cell produces two output pulses or slightly delayed output pulses with a clock pulse. Both of these cases are classified as logic "1" since, from a system-wide perspective, these cases do not produce errors.

Note that characterizing a locked OR cell at lower input currents is not necessary. Decreasing the input current in (2) exhibits the same effect as decreasing the coupling coefficient within a locked OR cell. For smaller input currents, the pattern remains the same. The key range, however, is constrained to a smaller region (see Fig. 14). For $I_{in} = 75\,\mu A$, the OR cell can be characterized within the dashed area at the center of Fig. 14.

For key values $K_1 = K_2 = 0.5$ and $I_{in} = 250\,\mu A$, a locked OR cell exhibits incorrect operation. Operation of the OR cell is shown in Fig. 15. In particular, the error occurs at approximately 500 ps, where the output is intended to be logic "1." The cell operates incorrectly due to a 2π phase shift of J_4 at around 400 ps. The 2π

phase shift occurs with the input combination of $In_1 = 0$ and $In_2 = 1$. Negative current therefore flows through inductor L_1. During the following clock pulse, the input combination of $In_1 = 1$ and $In_2 = 0$ restores the current through L_1 without producing a pulse at the output. Note that these explanations are common to all of the regions designated as "R" in Fig. 14. The input combination of $In_1 = 0$ and $In_2 = 1$ therefore destabilizes the current in the storage loop, making the output of the next clock cycle incorrect. The opposite input combination (i.e., $In_1 = 1$ and $In_2 = 0$) removes this effect and resets the OR cell back into normal operation. It is therefore possible to transform the "R" region into a region of correct operation by increasing the allowable space for the correct keys. The attack scenario is explained in the following subsection.

6.1.2 Attack Scenario

The proposed attack aims to *reset* the OR cell by applying the input combination $In_1 = 1$ and $In_2 = 0$ during each clock pulse. The "R" region depicted in Fig. 14 can be converted into the "C" region (i.e., correct operation of the OR cell). The correct key space therefore increases. A security parameter M is defined to quantify the ratio of the area of the correct operation region to the incorrect operation regions at $I_{in_{max}}$,

$$M \equiv \frac{\text{Correct key space}}{\text{Incorrect key space}}. \tag{4}$$

In the original locked OR cell [21], M is 1.59. By resetting the OR cell, the security parameter is increased to 2.69, corresponding to an 18.8% increase in the correct key space. The disadvantage of using this attack is lower throughput. For every input, a corresponding reset signal needs to be sent. The overall throughput therefore decreases by a factor of two.

In CMOS logic locking, threat models such as SAT attacks aim to decrease the incorrect key space, making a brute force search more efficient [47]. The reset threat model aims to increase the correct key space. From the perspective of the ratio of correct and incorrect key spaces, both of these threat models have a similar effect. The reset threat model can therefore be used in conjunction with other attacks to boost overall efficiency.

Although this attack focuses on a locked OR cell, a similar attack strategy for other types of locked circuits such as AND and XOR can also be applied. In the first step, the internal operation of a cell is analyzed to determine the specific input combination(s) that reset the cell into normal operation.

6.1.3 Possible Countermeasures

Limiting the key space prevents and thwarts the reset attack model. Selected coupling coefficients outside of the reset region (i.e., the "R" region in Fig. 14) limits

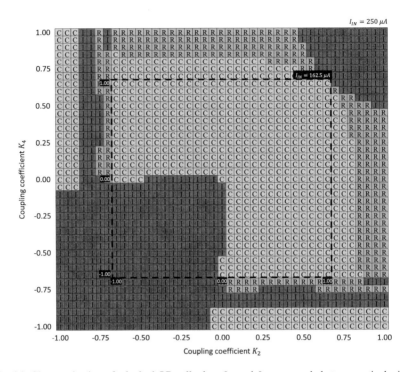

Fig. 16 Characterization of a locked OR cell when L_2 and L_4 are coupled at, respectively, input current $I_{in} = 250\,\mu A$ and $162.5\,\mu A$. The region notation is the same as shown in Fig. 14

the key space of the locked gate. However, choosing smaller coupling coefficients may not be an effective countermeasure due to the additional current flowing through the coupled inductor, as noted in (2). For example, in Fig. 14, the correct keys $K_1 = K_2 = 1$ at $I_{in} = 250\,\mu A$ can be transformed into the "R" region shown in Fig. 14 by decreasing the input current by two times (i.e., $I_{in} = 125\,\mu A$), which is equivalent to $K_1 = K_2 = 0.5$ at $I_{in} = 250\,\mu A$.

Another solution is to set a limit on the input current, which can be achieved by changing the line width and coupled inductors. By limiting the input current, the potential region is smaller. For example, for $I_{in_{max}} = 75\,\mu A$, the reset region ("R" region) is eliminated see Fig. 14 (note the dashed rectangular area at the center of the figure). In this case, $M = 1.04$.

The key space of the input current is significantly reduced by the limited input current. A different configuration of coupled inductors is therefore used; L_2 and L_4 as coupled inductors in the locked OR gate. This locked OR cell is characterized as shown in Fig. 16, where the correct key space can be increased by 32.9% by resetting the cell. In this case, $I_{in_{max}} = 162.5\,\mu A$ is sufficient to remove the possibility of resetting a cell, increasing parameter M to 2.88. A tradeoff therefore exists between M and $I_{in_{max}}$.

The bias margins of the SFQ OR cell are $\pm 30\%$ [48]. For the countermeasure of limiting the input current to be more robust against the reset attack model, the effect of process variations should be considered. The margins of a locked OR cell with coupled inductors L_1 and L_2 are evaluated. The margins of bias currents I_{b1}, I_{b2}, I_{b3}, and I_{b4} are, respectively, $(-45\%, +83\%)$, $(-36\%, +80\%)$, $(-82\%, +36\%)$, and $(-78\%, +34\%)$. The locked OR cell operates correctly under correct key values, and the reset attack model is not possible under any combination of coupling coefficients. The bias margins of the locked OR cell are therefore not degraded as compared to a standard SFQ cell. The margins of the coupled inductances, L_1, L_2, L_{M1}, and L_{M2}, are -68% and $+61\%$, much larger than the inductance variations in the MIT Lincoln Laboratory SFQ5ee fabrication process [49–51]. A margin analysis therefore supports robust operation of the locked cell when applying logic locking in SFQ systems.

6.2 Attack Model #2: Overproduction of Locked SFQ Circuits

The overproduction of complex locked circuits can be enabled by characterizing a locked cell and by sweeping the key values in polynomial time (i.e., the number of steps is linear rather than exponential). The threat model and attack scenario for overproduction based attacks are presented, respectively, in Sects. 6.2.1 and 6.2.2. Related countermeasures to thwart the proposed attack are described in Sect. 6.2.3.

6.2.1 Threat Model

In this threat model, an attacker is assumed to be located at the foundry, where the masks and layout of the fabricated device are available. The objective of the attacker is to produce a greater number of devices than requested by the company that developed the original IC design. Additionally, the attacker is assumed to know the type of logic locking technique (in this case, mutual inductance coupling). Since the device layout is known, correct operation of the circuit can be predicted with reverse engineering. In the case where the layout is protected from reverse engineering by camouflaged gates [22], a more powerful attack can be assumed given access to the simulation files.

In the overproduction threat model, the coupling coefficients are fixed to a specific (correct) value since the layout has previously been sent to the foundry, and no further modifications are possible once the device is fabricated. Although the coupling coefficients are fixed, the actual coefficients are difficult for an attacker to determine. The only key that remains under control is the input current. Due to pinout limitations in SFQ circuits, only one input current for the entire device is assumed. As a result, to successfully overproduce the device, the attacker should determine the correct range of input current.

A possible method to setup an attack is to apply a particular input combination, sweep the input current from 0 to $I_{in_{max}}$, and monitor whether the final output is correct. Although the attacker has access to the physical device, intermediate signals within the device cannot be monitored (i.e., only the output signals can be monitored). However, due to false positive outputs, the operation can be mistakenly classified as correct if a certain input combination is not applied to the locked cell. In the example of a locked OR cell, the key values within the reset region (the "R" region), shown in Figs. 14 and 16, can be considered as correct if the combination $In_1 = 0$ and $In_2 = 1$ is not evaluated. The attacker needs to check all of the input combinations for all of the input current values to ensure that the device operates correctly in all cases. For k number of inputs,

$$\sum_{i=1}^{2^k} \frac{I_{in_i}}{\Delta I_{in}} \tag{5}$$

measurements are required, where I_{in_i} is the range of input current that is swept during the ith iteration, and ΔI_{in} is the step size. (5) increases exponentially with a greater number of inputs (which is impractical in complex circuit designs). For example, assuming that $I_{in_i} = I_{in_{max}} = 250\ \mu A$ (i.e., worst case scenario) and $\Delta I_{in} = 25\ \mu A$, for $k = \{1, 2, 3\}$ number of inputs.{20, 40, 80}, measurements are required, which correspond to a 200% increase with one additional input signal.

6.2.2 Attack Scenario

In SFQ systems, the dependence of the current state of the output on a previous logic state is treated as a memory effect [52]. Alternatively, the operation performed during a clock cycle affects the operation during the subsequent clock cycle. Since the locked OR cell exhibits the memory effect with incorrect key values, an attacker can potentially apply an input sequence rather than just a single set of inputs to exploit this memory effect. An attacker needs to characterize a locked cell by setting $I_{in} = I_{in_{max}}$ and sweeping the coupling coefficients K_i from -1 to 1 with certain input sequences (similar to Fig. 14 without the reset region). This process can be described at the simulation level by either inferring the circuit configuration and parameters by reverse engineering the layout or directly accessing the simulation files (without the key values). The objective is to determine an input sequence that could reveal all possible incorrect key values of a locked OR cell. The input sequence should consist of all possible input combinations (e.g., for the 2 bit input, the combinations are {0,0}, {0,1}, {1,0}, and {1,1}) to trigger all of the internal states. Additionally, certain input combinations should be inserted at least two times to ensure the preceding input combination is different. This process is performed to trigger different variations of the memory effect. By evaluating different input combinations, an attacker can determine a particular input sequence to generate the same characterization plot as shown in Fig. 14. Note that this original characterization plot is generated with multiple input sequences rather than a single input sequence. For the locked OR cell with coupled L_1 and L_2 (see Fig. 11), the input sequence of

$$\{In_1, In_2\} = \{1, 1\} \rightarrow \{0, 0\} \rightarrow \{1, 0\} \rightarrow \{0, 1\} \rightarrow \{1, 0\} \rightarrow \{0, 0\} \qquad (6)$$

satisfies this condition. This input sequence is also shown in Fig. 15. Note that an overproduction attack is not only limited to this input sequence. The sequence in (6) is arbitrarily chosen. Other input sequences can also be used. The overproduction attack is also applicable for other locked cells (e.g., AND and XOR). Identifying the input sequence(s) that reveal(s) the incorrect operation of the circuit is required.

Once the input sequences are determined, those sequences should be applied to all locked cells within the device. By applying the input sequences for the first locked cell and monitoring the output, the range of input current can be determined that enable correct operation. The input current therefore converges to a certain range that corresponds to the correct key space. For a particular connection of cells, the locked cell may not receive a certain input combination. In this case, the attacker should characterize the cell under a limited range of input combinations and proceed with the attack. For example, if the locked OR cell (shown in Fig. 11) cannot receive the input combination $In_1 = 0$ and $In_2 = 1$, the "R" region shown in a new characterization should be in the "C" region shown in Fig. 14.

For the number of locked cells N in a circuit,

$$\sum_{i=1}^{N} \frac{I_{in_i}}{\Delta I_{in}} \qquad (7)$$

measurements are needed with the proposed overproduction attack. Equation (7) increases linearly with the number of locked cells and is independent of the number of inputs.

As a case study, a 4-to-2 priority encoder is treated as a circuit under attack. This priority encoder converts multiple input bits into a smaller number of output bits. The truth table of a 4-to-2 priority encoder is listed in Table 4 where V stands for the valid bit and X represents the don't care states. A 4-to-2 encoder is often used in interrupt controllers within processors to provide high priority interrupt requests [53]. This circuit also includes a considerable number of OR gates, which can be locked. The 4-to-2 priority encoder is therefore a useful topology to evaluate proposed attacks.

A 4-to-2 priority encoder is converted into SFQ by inserting delay elements, splitters, and logic gates [54]. The circuit is shown in Fig. 17. The coupling coefficients for four locked OR cells are specified in this figure. Note that the coupling coefficients of the locked OR cells are unknown to the attacker. Correct operation of this circuit is verified at $I_{in} = 250\,\mu A$, as shown in Fig. 18. The output signals are available four clock cycles after the inputs are applied.

Table 4 Truth table of a
4-to-2 priority encoder

In_3	In_2	In_1	In_0	Out_1	Out_0	V
0	0	0	0	X	X	0
0	0	0	1	0	0	1
0	0	1	X	0	1	1
0	1	X	X	1	0	1
1	X	X	X	1	1	1

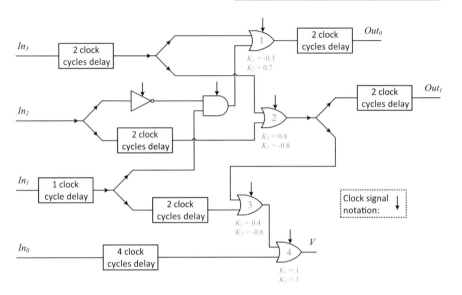

Fig. 17 4-to-2 priority encoder in SFQ

The maximum input current is assumed to be $250\,\mu A$ ($I_{in_{max}} = 250\,\mu A$). An overproduction attack is realized in four steps, where the number of steps is the same as the number of locked OR cells, see (7). The applied input sequences and expected (correct) output sequences are listed in Table 5. Each step corresponds to unlocking one of the OR cells labeled in Fig. 17. In Table 5, the input sequences are generated to produce (6) for each locked OR cell. The correct output sequence therefore always becomes an OR version of (6).

By generating the input sequences listed in Table 5 and monitoring the output signals, the attacker should record the range of input currents that produces correct operation. The results for each step are depicted in Fig. 19. To reduce the computational time, the range of correct input currents determined in one of the steps should be within a range of input currents in the following step (see Table 5).

In the last step, as shown in Fig. 19, the range of correct I_{in} is determined to be between 225 and $250\,\mu A$. In this range, the 4-to-2 priority encoder operates correctly under any input combinations and sequences. An overproduction attack is therefore successful on logic locked SFQ circuits.

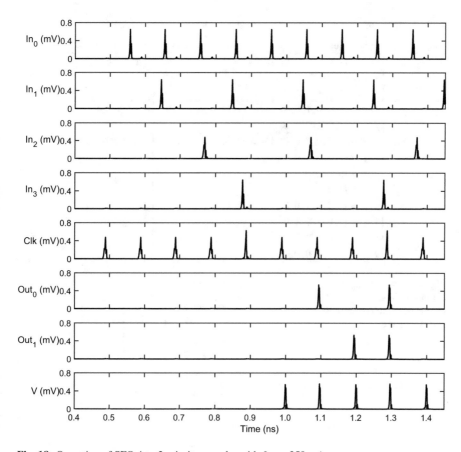

Fig. 18 Operation of SFQ 4-to-2 priority encoder with $I_{in} = 250\,\mu A$

6.2.3 Possible Countermeasures

A primary countermeasure against the overproduction threat model is to restrict the available knowledge of the attacker. In particular, by using camouflaged cells, deducing the correct operation of a device by reverse engineering the layout becomes significantly more difficult even for an experienced attacker. Since the correct operation is unknown in a camouflaged SFQ cell, the attacker cannot generate the desired input sequence for the locked cells. For an attacker that has access to the layout and simulation files except for the keys, a possible countermeasure is to increase the number of input current sources; two current sources, I_{in_1} and I_{in_2}, can be used rather than a single input current. The area overhead is, however, greater if an additional input current source is used to further increase the security of the circuit. A tradeoff therefore exists among the security of the system, area overhead, and number of input pins.

Table 5 Input and output sequences for each step in an overproduction attack

Step number	Input sequences	Correct output sequences	Range of input current sweep
1	$In_1 = \{1, 0, 0, 1, 0, 0\},$ $In_2 = \{0, 0, 1, 0, 0, 1\},$ $In_3 = \{1, 0, 1, 0, 1, 0\}$	$Out_0 = \{1, 0, 1, 1, 1, 0\}$	0 to 250 μA
2	$In_2 = \{1, 0, 0, 1, 0, 0\},$ $In_3 = \{1, 0, 1, 0, 1, 0\}$	$Out_1 = \{1, 0, 1, 1, 1, 0\}$	0 to 250 μA
3	$In_0 = \{0, 0, 0, 0, 0, 0\},$ $In_1 = \{1, 0, 0, 1, 0, 0\},$ $In_2 = \{0, 0, 0, 0, 0, 0\},$ $In_3 = \{1, 0, 1, 0, 1, 0\}$	$V = \{1, 0, 1, 1, 1, 0\}$	180 to 250 μA
4	$In_0 = \{1, 0, 0, 1, 0, 0\},$ $In_1 = \{1, 0, 1, 0, 1, 0\},$ $In_2 = \{0, 0, 0, 0, 0, 0\},$ $In_3 = \{0, 0, 0, 0, 0, 0\}$	$V = \{1, 0, 1, 1, 1, 0\}$	210 to 250 μA

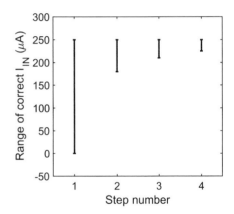

Fig. 19 Range of input current (I_{in}) resulting in correct operation of a 4-to-2 priority encoder

7 Conclusions

An important application of SFQ systems –large scale data centers operating with sensitive information– emphasizes the importance of hardware security in SFQ systems. Hardware security approaches for SFQ circuits –SFQ camouflaging and logic locking– are proposed herein. IC camouflaging and logic locking are well known techniques widely used to secure CMOS circuits. Although these techniques can be applied to SFQ circuits without modifications, standard approaches require a significant number of gates and additional input pins. IC camouflaging in SFQ circuits obstructs the reverse engineering process by inserting dummy JJs and camouflaged cells into a layout. A dummy JJ exhibits an identical top view image of a layout as a standard JJ. The rest of the layout and synthesis process remains unchanged. A large camouflaged SFQ circuit consists of camouflaged and regular gates with indistinguishable layouts. A novel method to provide a secret key for logic locking is also proposed. Standard SFQ gates can be modified to depend on a secret key current to maintain correct functionality. Mutual inductors can also be used to couple an additional positive or negative current to the locked gate from the key current. The efficacy of the proposed techniques is characterized by the number of camouflaged and logic locked gates. The area and power overhead of IC camouflaging and logic locking techniques are characterized with ISCAS'85 benchmark circuits. Tradeoffs among security, area, and power for these different approaches are evaluated. IC camouflaging increases the effort necessary for hardware-based reverse engineering attacks. Logic locking prevents an external attacker from analyzing the structural behavior of a design even if a copy of the secured circuit is obtained. Two new attacks models on logic locking, reset and overproduction, are evaluated. A 4-to-2 priority encoder is characterized to evaluate different attacks on logic locked circuits. Hardware security for superconductive computing systems can provide robust and trustworthy VLSI complexity SFQ circuits.

References

1. J. Bardeen, L.N. Cooper, R. Schrieffer, Theory of superconductivity. Phys. Rev. **108**(5), 1175–1204 (1957)
2. B.D. Josephson, Possible new effects in superconductive tunneling. Phys. Lett. **1**(7), 251–253 (1962)
3. T. Orlando, K. Delin, *Foundations of Applied Superconductivity* (Addison-Wesley, Boston, 1991)
4. T.V. Duzer, C.W. Turner, *Principles of Superconductive Devices and Circuits*, 2nd edn. (Pearson, London, 1981)
5. G. Krylov, E.G. Friedman, *Single Flux Quantum Integrated Circuit Design* (Springer, Berlin, 2022)
6. K.K. Likharev, V.K. Semenov, RSFQ logic/memory family: a new Josephson-Junction technology for sub-terahertz-clock-frequency digital systems. IEEE Trans. Appl. Supercond. **1**(1), 3–28 (1991)
7. H. Suhl, B.T. Matthias, L.R. Walker, Bardeen-cooper-schrieffer theory of superconductivity in the case of overlapping bands. Phys. Rev. Lett. **3**(12), 552–554 (1959)
8. T. Jabbari, G. Krylov, J. Kawa, E.G. Friedman, Splitter trees in single flux quantum circuits. IEEE Trans. Appl. Supercond. **31**(5) (2021)
9. T. Jabbari, J. Kawa, E.G. Friedman, H-tree clock synthesis in RSFQ circuits, in *Proceedings of the IEEE Baltic Electronics Conference* (2020)
10. T. Jabbari, E.G. Friedman, Global interconnects in VLSI complexity single flux quantum systems, in *Proceedings of the Workshop on System-Level Interconnect: Problems and Pathfinding Workshop* (2020), pp. 1–7
11. K. Gaj, Q.P. Herr, V. Adler, A. Krasniewski, E.G. Friedman, M. J. Feldman, Tools for the computer-aided design of multigigahertz superconducting digital circuits. IEEE Trans. Appl. Supercond. **9**(1), 18–38 (1999)
12. C.J. Fourie, Digital superconducting electronics design tools – status and roadmap. IEEE Trans. Appl. Supercond. **28**(5), 1–12 (2018)
13. G. Krylov, J. Kawa, E.G. Friedman, Design automation of superconductive digital circuits a review. IEEE Nanotechnol. Mag. **15**(6), 54–67 (2021)
14. V.K. Semenov, Y.A. Polyakov, S.K. Tolpygo, New AC-powered SFQ digital circuits. IEEE Trans. Appl. Supercond. **25**(3), 1–7 (2015)

15. T. Jabbari, G. Krylov, S. Whiteley, E. Mlinar, J. Kawa, E.G. Friedman, Interconnect routing for large-scale RSFQ circuits. IEEE Trans. Appl. Supercond. **29**(5) (2019)
16. T. Jabbari, E.G. Friedman, Flux mitigation in wide superconductive striplines. IEEE Trans. Appl. Supercond. **32**(3), 1–6 (2022)
17. T. Jabbari, G. Krylov, S. Whiteley, J. Kawa, E.G. Friedman, Repeater insertion in SFQ interconnect. IEEE Trans. Appl. Supercond. **30**(8) (2020)
18. T. Jabbari, E.G. Friedman, Surface inductance of superconductive Striplines. IEEE Transactions on Circuits and Systems II: Express Briefs **69**(6), 2952–2956 (2022)
19. T.V. Filippova, A. Sahua, A.F. Kirichenkoa, I.V. Vernika, M. Dorojevetsb, C.L. Ayalab, O.A. Mukhanov, 20 GHz operation of an asynchronous wave-pipelined RSFQ arithmetic-logic unit. Phys. Procedia **36**, 59–65 (2012)
20. J.Y. Kim, J.H. Kang, High frequency operation of a rapid single flux quantum arithmetic and logic unit. J. Korean Phys. Soc. **48**(5), 1004–1007 (2006)
21. T. Jabbari, G. Krylov, E.G. Friedman, Logic locking in single flux quantum circuits. IEEE Trans. Appl. Supercond. **31**(5) (2021)
22. H. Kumar, T. Jabbari, G. Krylov, K. Basu, E.G. Friedman, R. Karri, Toward increasing the difficulty of reverse engineering of RSFQ circuits. IEEE Trans. Appl. Supercond. **30**(3), 1–13 (2020)
23. J. Hurtarte, E. Wolsheimer, L. Tafoya, *Understanding Fabless IC Technology*. (Elsevier, Amsterdam, 2007), pp. 25–32
24. M.M. Tehranipoor, U. Guin, D. Forte, *Counterfeit Integrated Circuits*. (Springer, Berlin, 2015), pp. 15–36
25. T. Huffmire, B. Brotherton, T. Sherwood, R. Kastner, T. Levin, T.D. Nguyen, C. Irvine, Managing security in FPGA-based embedded systems. IEEE Des. Test Comput. **25**(6) (2008)
26. J.A. Roy, F. Koushanfar, I.L. Markov, EPIC: ending piracy of integrated circuits, in *Proceedings of the IEEE/ACM Design, Automation and Test Conference in Europe* (2008), pp. 1069–1074
27. S. Köse, L. Wang, R.F. DeMara, On-chip sensor circle distribution technique for real-time hardware trojan detection, in *Government Microcircuit Applications and Critical Technology Conference* (2017), pp. 1–4
28. M. Yasin, J.J. Rajendran, O. Sinanoglu, R. Karri, On improving the security of logic locking. IEEE Trans. Comput. Aided Des. Integr. Circuits Syst. **35**(9), 1411–1424 (2016)
29. J.A. Roy, F. Koushanfar, I.L. Markov, Ending piracy of integrated circuits. Computer **43**(10), 30–38 (2010)
30. P. Prinetto, G. Roascio, Hardware security, vulnerabilities, and attacks: a comprehensive taxonomy, in *Proceedings of the Italian Conference on Cybersecurity* (2010), pp. 177–189
31. W. Yu, O.A. Uzun, S. Köse, Leveraging on-chip voltage regulators as a countermeasure against side-channel attacks, in *Proceedings of the Design Automation Conference* (2015), pp. 1–6
32. W. Yu, S. Köse, False key-controlled aggressive voltage scaling: a countermeasure against LPA attacks. IEEE Trans. Comput. Aided Des. Integr. Circuits Syst. **36**(12), 2149–2153 (2017)
33. S. Seçkiner, S. Köse, Preprocessing of the physical leakage information to combine side-channel distinguishers. IEEE Trans. Very Large Scale Integr. Syst. **29**(12), 2052–2063 (2021)
34. A.W. Khan, T. Wanchoo, G. Mumcu, S. Kose, Implications of distributed on-chip power delivery on EM side-channel attacks, in *Proceedings of the IEEE International Conference on Computer Design* (2017), pp. 329–336
35. R. Torrance, D. James, The state-of-the-art in semiconductor reverse engineering, in *Proceedings of the ACM/EDAC/IEEE Design Automation Conference* (2011), pp. 333–338
36. J. Kunert, O. Brandel, S. Linzen, O. Wetzstein, H. Toepfer, T. Ortlepp, H. Meyer, Recent developments in superconductor digital electronics technology at FLUXONICS foundry. IEEE Trans. Appl. Supercond. **23**(5), 1,101,707–1,101,707 (2013)
37. J.F. Annett, *Superconductivity, Superfluids and Condensates* (Oxford University Press, Oxford, 2004)

38. F. Frost, R. Fechner, B. Ziberi, J. Völlner, D. Flamm, A. Schindler, Large area smoothing of surfaces by ion bombardment: fundamentals and applications. J. Phys. Condens. Matter **21**(22), 224026 (2009)
39. T. Kanayama, H. Tanoue, T. Tsurushima, Niobium silicide formation induced by Ar-ion bombardment. Appl. Phys. Lett. **35**(3), 222–224 (1979)
40. HYPRES Design Rules HYPRES, Inc (2015). https://www.hypres.com/wp-content/uploads/2010/11/DesignRules-6.pdf
41. T. Thangam, G. Gayathri, T. Madhubala, A novel logic locking technique for hardware security, in *Proceedings of the IEEE International Conference on Electrical, Instrumentation and Communication Engineering* (2017), pp. 1–7
42. G. Krylov, E.G. Friedman, Design methodology for distributed large-scale ERSFQ bias networks. IEEE Trans. Very Large Scale Integr. Syst. **28**(11), 2438–2447 (2020)
43. F. Brglez, H. Fujiwara, A neutral netlist of 10 combinational benchmark circuits, in *Proceedings of the IEEE International Symposium on Circuits and Systems* (1985), pp. 685–698
44. P. Subramanyan, S. Ray, S. Malik, Evaluating the security of logic encryption algorithms, in *Proceedings of the IEEE International Symposium on Hardware Oriented Security and Trust* (2015), pp. 137–143
45. Y. Mustafa, T. Jabbari, S. Köse, Emerging attacks on logic locking in SFQ circuits and related countermeasures. IEEE Trans. Appl. Supercond. **32**(3), 1–8 (2022)
46. F.A.P. Petitcolas, *Kerckhoffs' Principle* (Springer US, New York City, 2011). https://doi.org/10.1007/978-1-4419-5906-5_487
47. R. Hofstede, M. Jonker, A. Sperotto, A. Pras, Flow-based web application brute-force attack and compromise detection. J. Netw. Syst. Manag. **25**, 735–758 (2017)
48. S.V. Polonsky, V.K. Semenov, P.I. Bunyk, A.F. Kirichenko, A.Y. Kidiyarova-Shevchenko, O.A. Mukhanov, P.N. Shevchenko, D.F. Schneider, D.Y. Zinoviev, K.K. Likharev, New RSFQ circuits (Josephson Junction digital devices). IEEE Trans. Appl. Supercond. **3**(1), 2566–2577 (1993)
49. S.K. Tolpygo, V. Bolkhovsky, T.J. Weir, C.J. Galbraith, L.M. Johnson, M.A. Gouker, V.K. Semenov, Inductance of circuit structures for MIT LL superconductor electronics fabrication process with 8 niobium layers. IEEE Trans. Appl. Supercond. **25**(3), 1–5 (2015)
50. S.K. Tolpygo, V. Bolkhovsky, T.J. Weir, A. Wynn, D.E. Oates, L.M. Johnson, M.A. Gouker, Advanced fabrication processes for superconducting very large-scale integrated circuits. IEEE Trans. Appl. Supercond. **26**(3), 1–10 (2016)
51. S.K. Tolpygo, E.B. Golden, T.J. Weir, V. Bolkhovsky, Inductance of superconductor integrated circuit features with sizes down to 120 nm. Supercond. Sci. Technol. **34**(8), 1–24 (2021)
52. T. Ortlepp, M.H. Volkmann, Y. Yamanashi, Memory effect in balanced Josephson comparators. Phys. C **500**, 20–24 (2014)
53. Intel, *8259A Programmable Interrupt Controller* (8259A/8259A-2). (Intel Corporation, Santa Clara, 1988)
54. M.D. Ciletti, M.M. Mano, *Digital Design* (Prentice-Hall, Hoboken, 2007)

Index

A
Application-specific architecture, 42, 48, 57, 58, 65
Architecture design, v, 38, 41–65
Automatic niching particle swarm optimization (ANPSO), 106, 107, 120–128, 130, 131

C
Camouflaging, 137–139, 162, 163
Computer Aided Design (CAD), 88, 90–92
Countermeasures, 147, 153, 155–157, 161–162

D
Decision diagrams, v, 1–21
Decoherence, 11, 12, 14, 74–75
Design automation, vi, 1, 21, 28

F
Field programmable gate array (FPGA), 87–102
Fireworks algorithm (FWA), 106, 107, 120–128, 130, 131

H
Hardware security, v, 135–163

L
Logic cells, 105–131, 141–142
Logic locking, 5, 137, 138, 146–163

M
Margin, 105–131

N
Noisy intermediate-scale quantum (NISQ), 1

O
Optimization methods, 108, 112–117, 121

Q
Qiskit, 31, 36, 57, 74
Quantum circuit equivalence checking, 15–17
Quantum circuit simulation, 2, 11, 12, 14, 15
Quantum computer, v, vi, 7, 11, 12, 21, 25–27, 42, 69, 70, 74, 76, 77, 83–85
Quantum computing (QC), 1–21, 25–38, 41
Quantum gate error, 11, 12, 65, 69, 70, 73–75, 77, 79, 84
Quantum optimization, 72, 77
Quantum readout error, 69, 72–75, 77–79, 84
Quantum scheduling, 30, 33
Quantum state tomography, 70, 71
Quantum true random number generator, 69–85
Qubit mapping, 27, 28, 31–35, 45–46, 59, 61, 64, 65
Qubit placement, 45, 50, 51, 63

© The Author(s), under exclusive license to Springer Nature Switzerland AG 2023
R. O. Topaloglu (ed.), *Design Automation of Quantum Computers*,
https://doi.org/10.1007/978-3-031-15699-1

R
Rapid Single Flux Quantum (RSFQ), 87–94,
    100–102, 106, 109, 112, 113, 115, 119,
    121, 131
Reverse engineering (RE), 137, 138, 147, 157,
    158, 161–163

S
Single flux quantum, 105–131, 136

Single flux quantum logic, 105–131
Superconducting electronics (SCE), 87,
    88, 102
Superconducting quantum circuit, 41,
    44–47
Superconductive digital electronics, v, 136
Superconductive integrated circuits (ICs), 147
Superconductor electronics, 135, 137

Printed in the United States
by Baker & Taylor Publisher Services